Understanding The Origin and Diversity of Life

by

Elizabeth J. Ridlon

and

Robert W. Ridlon, Jr.

Understanding the Origin and Diversity of
Life

Copyright © 1997
Elizabeth J. Ridlon and Robert W. Ridlon, Jr.
First Printing 1998
ISBN: 1-57502-719-4

Scripture references are from a New International Version (NIV) and are copied from the macBible™ Software for Bible Study and Research. Copyright © 1990, Zondervan Electronic Publishing, Grand Rapids.

All photographs were taken by the authors.

Published by: Jordan Hall Publishing
                  P.O. Box 364
                  Troy, Illinois 62294-0364

Cover photograph, taken by Robert W. Ridlon, Jr., is a fossil fish identified as *Diplomystus* from the Green River Formation, Kemmerer, Wyoming.

Printed in the USA by

MP
MORRIS PUBLISHING

3212 East Highway 30 • Kearney, NE 68847 • 1-800-650-7888

Preface

As we will point out in the first chapter, science and the Bible are fully compatible. Through the various disciplines of science, the creation is revealed from the microcosm of chemistry to the macrocosm of the stars and planets. The Apostle Paul writes in his letter to the church in Rome that the creation is a witness of God. Surely if God has creation as a witness of Himself, we ought to have an understanding and appreciation of that witness. It's also interesting that many people feel that there is an incompatibility between believing science and believing the Bible. When Elizabeth was teaching college biology, one of her students was very interested in how one could be both a Christian and a scientist. Elizabeth was able to share her faith in Christ and at the same time explain that science was compatible with the Bible.

Many people blindly accept the ideas of evolution and some even believe that God may have used evolution as His way of bringing about the creation. Unfortunately, this point of view (called theistic evolution) sounds good, but when the real premises of evolution are known, it becomes obvious that there are some major problems with that concept.

Purpose

The purpose in writing this book, is to give the reader an understanding of the building blocks that are used in both the creation model and the evolution model. By providing some basic chemistry and biology, it is much easier to understand and critically view either model.

This book doesn't examine every point of the evolution model and refute it point by point. The basic principles of evolution are presented and addressed to show some pertinent serious flaws. Although we don't present all the arguments for the creation model, we will present a scientific view that is supported by the scriptures. The focus of this book is on the living world with support from the physical sciences.

Format and Organization

Our vision is for the reader to be able to use this book as a guided self study or use it in group training led by an instructor. The book is divided into 7 chapters, each of which is designed to be studied in 45 to 60 minutes including discussion. Some thought provoking questions are posed at the end of each chapter. Chapter 1 introduces the concepts of creation and evolution. Chapter 2 establishes science as valid and explains the scientific method. Chapters 3 and 4 develop the reader's understanding of the building blocks of life (i.e., chemistry and biology). Chapter 5 presents and explains the creation model from the scriptures in Genesis. Chapter 6 presents and critically examines some of the main points of the evolution model. Chapter 7 summarizes the salient conclusions of the previous chapters and addresses some odds and ends, such as the age of the earth and dinosaurs.

At the end of the book, there is a reading list and bibliography. Quoted material is always cited and included in the bibliography. Some of the material is referenced in order to give citation to the original authors. For example, in Chapter 7, there is a section on fossils. In that section we have included a summary of six processes described in a book by MacFall and Wollin. In this case, the citation was given since their book was source of the material. Sometimes the material covered is of a general nature and in these cases no direct citation is made; however, a general reference is provided in the reading list to allow readers to gain further information on that subject. For example, most of Chapter 4 (The Principles of Biology) is considered common knowledge and just about any modern biology textbook would contain these principles. Therefore, we chose several biology texts and included them in the bibliography. All scripture references are from a New International Version (NIV) and are copied from the macBible^{TM} Software for Bible Study and Research from Zondervan Electronic Publishing, Grand Rapids, 1990.

Our Commitment

We have been involved in science for over twenty years. We believe very strongly that, through science, God has much to say about His creation and His Plan. As Christians we want the world to know there is a God and that He loves us and wants to be with us forever. Through His plan of salvation -- Jesus dying on the cross to pay for our sins -- that is possible. Unfortunately there are veils of uncertainty that try to get in the way of many so they don't see the Truth. Paul wrote to the Corinthians that:

> 3 And even if our gospel is veiled, it is veiled to those who are perishing. 4 The god of this age has blinded the minds of unbelievers, so that they cannot see the light of the gospel of the glory of Christ, who is the image of God.
> (NIV 2 Corinthians 4: 3-4)

Robert W. Ridlon, Jr.
Elizabeth J. Ridlon

1997

Table of Contents

Chapter 1

Introduction

Roger Oakland, a creationist lecturer said "If there is one thing that has caused people to question the reality of God the Creator more than anything else, it is the teaching of evolution." Michael Denton wrote in his book *Evolution: A Theory in Crisis*, that the question of evolution is being debated today at an intensity that is as great as when the theory was first debated in the 19th century.

Many scientists have questioned the evolution model and recognize that it is full of problems and fundamentally wrong. Other scientists have defended the theory vehemently. Just about every issue of *National Geographic* magazine carries some reference or inference to evolution. Many view this as a desperate attempt to gain support for evolution as it may be headed for more problems.

Rejecting evolution goes beyond the mere rejection of a theory for many scientists. If the creation model is accepted as true, this is acknowledging that there is a Creator and that being the case, there is a Divine order and we are all part of it. It adds a dimension of being responsible to our Creator for our actions. This may be one of the reasons so many have clung tightly to evolution -- it takes God out of the picture.

The Creation Model

The basic premise of the Creation model is found in Genesis 1:1 which states that" In the beginning God Created the heavens and the earth. "

The Apostle Paul restates the belief in his letter to the Colossians stating:

> For by him all things were created: things in heaven and on earth, visible and invisible, whether thrones or powers or rulers or authorities; all things were created by him and for him. (NIV Colossians. 1:16)

Beginning with Chapter 5 of this book, we will look at the details of the Creation model.

There are two main principles that can be derived from the creation model. The first is that life was created through an intentional act of God -- not by chance. Second, the diversity (or differences) among organisms is attributable to God's original creation. Although there is variability within species (e.g., poodles and collies), this doesn't give rise to new species (e.g., collies don't become bears). We will see in Chapter 2, that as compelling as the arguments are for creation or seem to be for evolution, neither can be proven scientifically. The matrix shown in Figure 1 illustrates the basic differences between the creation model and the evolution model.

The Evolution Model

In his book *Evolution: Challenge of the Fossil Record*, Duane Gish summarizes the definition of evolution, based on Darwin's General Theory of Evolution, as: "all living things have arisen by a naturalistic, mechanistic, evolutionary process from a single living source which itself arose by a similar process from a dead, inanimate world" (p. 28). Evolution allows for and explains the beginning of new kinds of organisms; even the beginning of the very first living organism. We usually attribute the concepts of evolution to Charles Darwin, who published a book on his theory in 1859. However, Darwin didn't have the original idea.

The concept that living things originated as a result of chance probably had its earliest origins from the classical Greek philosophers, such as Anaximander (550 BC) and Empedocles (450 BC).

	Origin of Life	Diversity of Life
Creation Model	Created by God	Different "kinds" created by God. Wide variation within species is possible.
Evolution Model	Chance assemblage of particles	Different "kinds" result from natural processes favoring mutations. Extreme variations within a species may give rise to new species.

Figure 1. A comparison of the creation and evolution models

4

There are two basic principles of evolution. The first is that nonliving material becomes living material. This is a basic premise since the idea of a Creator (God) is dismissed and instead life was generated by some set of circumstances.

The second principle is that the diversity of life is explained by natural processes such as natural selection and mutations. The evolution model says mutations improve life and cause it to evolve to more complex organisms. Interestingly, the concept of natural selection is in agreement with both the Bible and science, but natural selection doesn't give rise to new species, it merely allows changes to take place in the physical form of a species to give it a better survival opportunity. We will look at this more in depth in Chapter 6, but for now the bottom line is that there is no evidence that one species evolved from another.

What is Wrong With Evolution?

It's interesting that every year scientists learn more about the problems associated with the evolution model, yet many still cling to it. Why? It may be that as we have suggested, the concept of evolution is more than just a model, but rather it serves a greater purpose of being antagonistic toward the truth of a Divine creation.

Many researchers believe it is really another religion and parts of old religions in disguise. There is evidence that supports the connection between evolution and most of the pagan religions. Erasmus Darwin, the great grandfather of Charles Darwin, was head of the Lunar Society in London and wanted to explain away God. Karl Marx, a contemporary of Charles Darwin, believed evolution to be a pillar of communism.

An extensive research conducted by Bergman reveals that Hitler, as well as his top aids, believed and embraced the ideas of evolution and in particular the idea of survival of the fittest. Bergman points out that the complete title of Darwin's work was *The Origin of Species by Means of Natural Selection or the Preservation of Favored Races in the Struggle for Life.*

A final thought is that when Genesis is put into question, then the rest of the Bible is put into question. Our faith in God as Creator is certainly important as pointed out by the writer of Hebrews:

> 1   Now faith is being sure of what we hope for and certain of what we do not see. 2   This is what the ancients were commended for. 3   By faith we understand that the universe was formed at God's command, so that what is seen was not made out of what was visible. (NIV Hebrews 11: 1-3)

In his letter to the Romans, the Apostle Paul makes a strong point regarding the suppression of Truth:

> 18 The wrath of God is being revealed from heaven against all the godlessness and wickedness of men who suppress the truth by their wickedness, 19 since what may be known about God is plain to them, because God has made it plain to them. 20 For since the creation of the world God's invisible qualities --his eternal power and divine nature --have been clearly seen, being understood from what has been made, so that men are without excuse. 21 For although they knew God, they neither glorified him as God nor gave thanks to him, but their thinking became futile and their foolish hearts were darkened. (NIV Romans 1:18-21)

Verse 21 could refer to some scientists. Romans 1 verse 20 is a powerful warning from God since it states that because of creation no one has an excuse to be a nonbeliever. God has made himself known through creation.

Conclusion

As we guide a study of the creation model and examine the problems with evolution model, we hope to accomplish three things. First, we think that science can be used as a witness of God as it says in Romans 1:20. Second, we will explain the foundations of the creation model from a scientific perspective. Christians should be able to appreciate science as a way to see and better understand God's Creation. In Genesis Chapter 1, we see God at work creating the land, seas, plants, sun, moon, stars, animals, and especially man. Throughout the process God saw that it was good. Third, we will critically examine the evolution model.

Discussion Questions

1. Why do you think some scientists stick to evolution even when there is a lack of evidence to support it?

2. Do you think it is important to decide between the creation model and the evolution model? Why?

Chapter 2

Science and the Scientific Method

Daniel Vestal in his book, *Doctrine of Creation*, says "God has revealed Himself in Scriptures as the God who acts in history. The created order is the arena in which He is working out His unfolding plan. This plan is being worked out within the boundaries of time and space" (p. 62). As we pointed out in Chapter 1, the creation is an important witness of God.

Prior to 1859 and the evolution theory, science was not viewed as antagonistic to Christianity. Science was actually a friend of religion and was viewed as providing continuing support for the idea of wisdom of a creator and the grandeur of His design. The introduction to the first issue of the prestigious *Zoological Journal of London*, published in 1824, made this statement: "The naturalist...sees the beautiful connection that subsists throughout the whole scheme of animated nature. He feels too that at the head of all this system of order and beauty, preeminent in the domain of his reason, stands Man...the favoured creature of his Creator" (Quoted in Denton, p. 20).

In 1857, one of the leading biologists of North America, Louis Agassiz, Professor of Zoology at Harvard, wrote that the living world "shows also premeditation, wisdom, greatness, prescience, omniscience, providence...all these facts proclaim aloud the One God whom man may know" (Quoted in Denton, p. 20).

Henry Morris wrote a book called *Men of Science, Men of God*. In that book, he points out that there have been thousands of scientists that were or are Bible-believing Christians. Some of the notable examples in the book are: Leonardo da Vinci, Blaise Pascal, Isaac Newton, Charles Babbage, and Louis Pasteur. Duane Gish (a PhD from the University of California in Biochemistry) believes that the evolutionists would like everyone to think that scientists don't believe in creation.

Russell Humphreys, a physicist at Sandia National Laboratories in New Mexico said in a 1996 article that he estimates "there are at least 10,000 practicing scientists in the U. S. who are young earth creationists" (McCulley, pp. 31-32). Humphreys is a member of the Creation Research Society which has over 3,000 members and 600 voting members. These are practicing scientists with graduate degrees and include physicists, geologists, biologists, chemists, medical doctors, engineers, and mathematicians (McCulley).

In this chapter, we will look at what science is and what makes up the scientific method. This will form a foundation which will make the concepts of creation model more understandable. This foundation will provide a basis for critically examining the evolution model as well.

Science

Science is used to study the laws of the physical (non-living) and biological (living) world. There are different divisions called disciplines within science. Two of these disciplines are the life sciences (biological sciences) and physical sciences.

Some areas of study within the biological sciences include:

Discipline	Area of Study
Botany	Plants
Zoology	Animals
Ethology	Animal Behavior
Ecology	Organisms & Environment
Entomology	Insects
Herpetology	Reptiles
Mammalogy	Mammals
Ichthyology	Fish
Ornithology	Birds

The physical sciences are disciplines concerned with the sciences of matter and energy.

11

Some areas of study within the physical sciences are:

Discipline	Area of Study
Chemistry	Matter
Physics	Energy and Matter
Astronomy	Planets and Stars
Geology	Rocks and Minerals
Meteorology	Weather

Science has illuminated many mysteries of creation. Many scientific discoveries have gone far to improve our quality of life. Every day, new discoveries are made in the disciplines of biology, medicine, and chemistry that save lives and improve our health. We owe much to the men and women who endeavor to understand our world and make it a better place to live. Many of these scientists realize that God created the universe and it is God that allows these discoveries to be made.

The Scientific Method

Scientists use the scientific method to study the biological and physical world. Personal opinions, family traditions, an explanation of a situation from friends, or cultural traditions are not part of the scientific method. There are six basic steps that comprise the scientific method.

Step 1 - *Identify the Problem*
Identify the problem or phenomenon to be investigated. Generally a scientist

12

already has accumulated many facts and made observations about the problem.

Step 2 - *Formulate an Hypothesis*
An hypothesis is an educated guess or explanation of the problem and is often stated as an IF - THEN statement. An example is : IF red food dye is fed to mice, THEN they will develop stomach cancer. The hypothesis must be testable in an experiment. Another way to state the hypothesis is as a null hypothesis. The null hypothesis is an hypothesis that is written in the negative. For example: Red food dye does not cause stomach cancer in mice. If during experimentation, stomach cancer is found in the mice getting the red dye, then the hypothesis is proved wrong (rejected) and one can say red food dye *might* cause cancer in mice.

Step 3 - *Test the Hypothesis*
The hypothesis is tested by an experiment where data is collected and then analyzed. The experiment will involve observations of some type. Usually the observations will be measuring or counting something such as how many birds at a feeding station. These observations are known as data, which can be mathematically analyzed. Very often experiments involve two groups. One group is the control group and nothing experimental is done to this group -- but it is needed to see if an unknown factor could be causing some phenomenon to occur. The other group is called the experimental group and this is the group that is experimentally tested. From the previous example, one

13

group of mice could be fed red food dye in their food to see if they get stomach cancer. This is the experimental group. Another group of mice is fed the same food without the red dye and is, in every way possible, treated the same as the experimental group. These mice are called the control group. It is called the control group because the scientists want to make sure it is the dye and not something else that causes the cancer. If the control group were to get cancer, then something else other than the red food dye may be the cause.

Step 4 - *Analyze the Data and Formulate Conclusions*
This is actually the hardest part of the scientific method. Although the data analysis is presented objectively, there may be some speculation or opinion involved in the interpretation. Usually the experiment will either support or reject the hypothesis.

Step 5 - *Report the Data in a Scientific Journal*
This is important so others can read and think about the experiment. The report includes the hypothesis and details of how the experiment was conducted, as well as the results.

Step 6 - *Verify the Hypothesis*
Other scientists should be able to duplicate the experiment and get the same results. This serves to verify that what was found in the experiment was valid.

Theories

Much is written concerning evolution and the so-called "theory of evolution." We need to understand what a theory is before we can apply it appropriately to either evolution or creation. The word *theory*, as used in the scientific world, has a little bit different meaning than the word used in everyday English. Usually *theory* in everyday conversation has about the same meaning as hypothesis -- such as "I have a theory about how our neighbor's house burned down -- it might have something to do with their child loving to play with matches." In scientific language, however, a theory is a very well tested hypothesis and it is not just a trivial hypothesis such as "vitamins can cure cancer." Theories are broad generalizations about how the universe works including the biological world. Some examples of currently accepted theories in biology include:

> *Cell Theory.* All organisms are composed of one or more cells with the exception of viruses which are only made of a protein coat and either DNA or RNA. The cell is the structural unit of all organisms. All cells come from preexisting cells. (Incidental, the evolutionist must affirm that there was one exception when the first cell came from a conglomeration of chemicals.)

> *Biogenesis Theory.* Life comes only from life. In other words no person or

15

scientist has ever created life, even small forms such as bacteria.

DNA/RNA is the Blueprint for Life. All life contains DNA/RNA (found in chromosomes) which is coded information that dictates or controls what the animal/plant will be like: its structure or anatomy, its function or physiology and its behavior.

A theory provides a good explanation for many scientific observations. Perhaps most importantly, a theory has been tested by many scientists over a long period of time. It has withstood the test of time and found to be true within the realm of our present human knowledge. A theory is also a good tool for predicting what will happen in the physical and biological world.

Scientists and laymen may even think and talk about theories as if they were facts; however, a theory can still be proved wrong. In 140 AD a Greek astronomer Ptolemy put forth the geocentric theory which stated that the sun revolves around the earth. It was not until 1580 AD that Copernicus presented the heliocentric theory that the earth revolves around the sun. His theory better explained and fit the observations (Chiras).

Theories must meet three criteria: Its events, processes, or properties must be (1) observed, (2) useful in predicting future events in nature or experimentally, and (3) capable of being subject to disproof (Gish). Both the evolution "theory" and creation "theory" fail to meet these criteria. Therefore, we will have to be satisfied with calling them both models. In order to evaluate these models, observations can be made and there should be some evidence that fits the predictions of each one. In Chapters 5 and 6, we will see that there are in fact very definite expectations on the parts of both models that can be critically examined.

In the next chapter, we will look at some very basic chemistry so we can understand some biochemical principles that are used in both the creation and evolution models.

Discussion Questions

1. Do you think it's easy or difficult to be a scientist and be a Christian?
Why?

2. How would you respond to the statement that evolution is a fact?

Chapter 3

The Principles of Chemistry and Biochemistry

Chemistry is the study of the composition, properties, and behavior of the material world. Chemists study how substances are formed and learn to predict under what conditions substances will react with each other. Chemistry is also concerned with the states of matter (i.e., solid, liquid, or gas). The world around us is comprised of various objects. These objects are made of materials that may be pure (such as gold) or may be combinations of various substances such as plastic. Even water is a combination of two substances: hydrogen and oxygen.

In order to understand the creation and to be able to understand the claims of the evolution model, it is important to have a basic understanding of chemistry. Understanding the origin of life, the age of the earth, and organization of living things depends on an understanding of chemistry. We will consider some of the radiometric dating techniques applied by evolutionists to substantiate their claim of an old earth. Our study of chemistry will allow us to understand the methods and the shortcomings of these techniques. In this chapter we will begin by explaining atoms, molecules, and compounds; and then organic compounds found in living organisms.

Atoms and Elements

Atoms are the smallest unit of matter. One cannot split or divide an atom by normal chemical means. However, under special conditions the atom can be split releasing tremendous amounts of energy (e.g., the atomic bomb or a nuclear power plant). There are over 100 different kinds of atoms which we call elements. An element is a pure substance that is made of one kind of atom. For example, oxygen atoms, hydrogen atoms, gold atoms, silver atoms, and carbon atoms are a few of the elements.

All atoms have a central nucleus with electrons surrounding it. The central nucleus contains protons and neutrons. Each proton has a positive electrical charge. Neutrons have no electrical charge. Neutrons and protons are very small but very heavy. 1/3 teaspoon of neutrons and protons from any element that are packed together would weigh 100 million tons. However, we know that 1/3 teaspoon of an element does not weigh that much. Even 1/3 teaspoon of gold, which is very heavy, doesn't weigh much. Why?

The answer is because there is a lot of empty space in a teaspoon of gold. The space occurs between the nucleus and electrons.

Electrons have a negligible weight and a negative electrical charge. They are arranged in layers (called orbitals) and revolve around the nucleus. Electrons with the least energy are in orbitals closest to the nucleus. Electrons with the most energy are in orbitals farther away from the nucleus.

The structure of the helium atom, which has only two electrons is represented by the drawing in Figure 2.

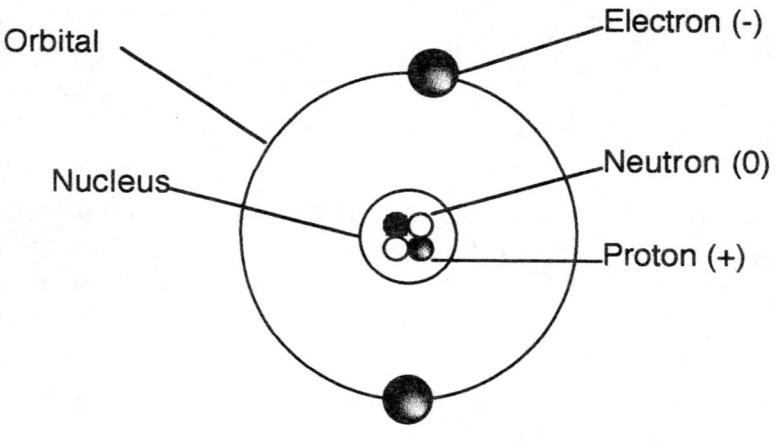

Figure 2. A Helium Atom

Elements are often abbreviated by letters (e.g., C for carbon; O for oxygen). Also, elements are numbered according to the number of protons they contain. We start with number 1 and continue to more than 100. This number is called the *atomic number*. The atomic number equals the number of protons. The atomic number also equals the number of electrons in a neutral atom. Some examples:

Hydrogen has the atomic number 1; it has 1 proton and 1 electron.

Oxygen has the atomic number 8; it has 8 protons and 8 electrons.

Nitrogen has the atomic number 7; it has 7 protons and 7 electrons.

Sodium has the atomic number 11; it has 11 protons and 11 electrons.

The *atomic weight* equals the number of protons plus the number of neutrons. Protons and neutrons are responsible for the weight of an atom and as we said before, electrons are practically weightless. Examples of atomic weight:

Carbon has atomic weight of 12 and atomic number is 6. Atomic wt. (12) = protons (6) + neutrons (6).

Hydrogen has atomic weight of 1 and atomic number of 1. Atomic wt (1) = protons(1) + neutrons (0).

21

Oxygen has atomic weight of 16 and atomic number of 8. Atomic wt (16) = protons (8) + neutrons (8.)

Nitrogen has atomic weight of 14 and atomic number of 7. Atomic wt (14) = protons (7) + neutrons (7).

Sodium has atomic weight of 23 and atomic number of 11. Atomic wt (23) = protons (11) + neutrons (12).

The following chart puts it all together.

Element	Abb	Atomic #	Atomic Wt	Protons	Neutrons	Electrons
Carbon	C	6	12	6	6	6
Hydrogen	H	1	1	1	0	1
Oxygen	O	8	16	8	8	8
Nitrogen	N	7	14	7	7	7
Uranium	U	92	238	92	146	92
Lead	Pb	82	207	82	125	82
Potassium	K	19	39	19	20	19

Isotopes

Sometimes elements have a variable number of neutrons in their nucleus. When this occurs, the element is said to have isotopes. For example, carbon in its most common form has an atomic weight of 12 (6 protons and 6 neutrons). However, there are small quantities of carbon 13 (an isotope of carbon) which has 7 neutrons. Another isotope occurs as carbon 14 which has 8 neutrons. All three are known isotopes of carbon.

Another example is hydrogen which has three isotopes. Hydrogen 1 (protium), is the most common and has one proton and no neutrons. Hydrogen 2 (also called deuterium) has one proton and one neutron. Finally, hydrogen 3 (or tritium) has one proton and two neutrons.

An isotope that decomposes (nucleus changes) spontaneously is called a *radioactive isotope.* Radiation is emitted by radioactive isotopes and can be dangerous or fatal in high amounts.

The decomposing isotope gains or loses protons and may also gain neutrons. As a result, the radioactive isotope becomes a different element called a daughter product. Some isotopes break down very quickly and some take many years. Two important isotopes of uranium are uranium 235 and uranium 238. Both the uranium 235 and the 238 isotopes break down to lead. For example, U-238 begins with 92 protons and 146 neutrons. Through a series of complex steps over time, it loses 10 protons and 22 neutrons, leaving it with only 206 nuclear particles which is now the lead isotope (Pb-206). The time its takes for 50% of the original U-238 to become Pb-206 is called its halflife. The halflife of uranium 238 is 4.5 billion years.

The application to dating rocks is based on knowing the amount of uranium and lead in a given sample. If a rock originally contained ten grams of uranium, it is presumed that after 4.5

23

billion   years there would be five grams of uranium and five grams of the daughter product lead formed.   Obviously, certain assumptions must be made.   First, the relative amounts of original uraniumm and lead must be known. Also, it is assumed that the decay rate has not changed since the formation of the original rock. Third, nothing has occurred over time to change the amounts present for either element.   These are three assumptions that bring the dating process, used by evolutionists, into question.

Compounds and Molecules

A *compound* is a substance that contains two or more different elements.   For example, water contains the elements hydrogen and oxygen. Glucose contains hydrogen, oxygen, and carbon. Carbon dioxide contains the elements of carbon and oxygen.

A *molecule* contains two or more atoms.   A molecule is the smallest amount of a compound that can occur.   For example a molecule of water contains two atoms of hydrogen and one atom of oxygen and is written H_2O.   A molecule of glucose contains six atoms of carbon, six atoms of oxygen and 12 atoms of hydrogen and is written $C_6H_{12}O_6$.   Oxygen is a gas and occurs as a molecule containing two atoms of oxygen (i.e., O_2).

Figure 3 shows drawings of water, oxygen, and carbon dioxide molecules.

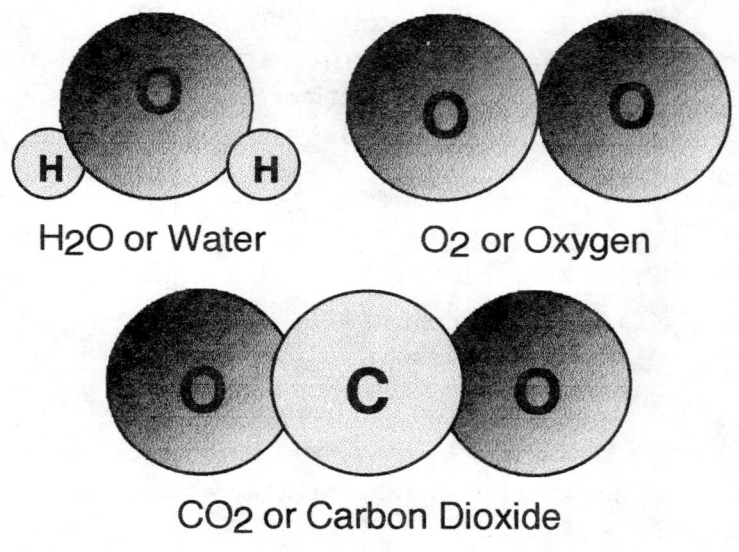

H₂O or Water

O₂ or Oxygen

CO₂ or Carbon Dioxide

Figure 3.   Molecules

There are two types of molecules: those that contain the carbon atom and are called *organic molecules* (e.g., glucose) and those that do not contain the carbon atom and are called *inorganic molecules* (e.g., water). There are four large organic compounds that make up living things. These are *proteins, carbohydrates, lipids (fats)* and *nucleic acids.*

Proteins are polymers (chains) of amino acids. They are found in the skin, hair, body tissues (e.g., muscle), and enzymes. Amino acids are chemically strung together (like beads on a bracelet) to form a protein. There are 20 basic amino acids found in plants and animals, and each one has an abbreviation such as Ala for alanine. The amino acid general structure is shown in Figure 4.

```
                                          H
  NH3 = amino group                       |
  COOH = carboxylic acid          NH3--C--COOH
  R = remainder of molecule                |
  which varies                             R
```

Figure 4.   Amino Acid Structure.

The 20 different amino acids have different R groups. Here is a list of the 20 different amino acids (and their abbreviations) that make up all living things:

Alanine	Ala	Leucine	Leu
Arginine	Arg	Lysine	Lys
Asparagine	Asn	Methionine	Met
Aspartate	Asp	Phenylalanine	Phe
Cysteine	Cys	Proline	Pro
Glutamate	Glu	Serine	Ser
Glutamine	Gln	Threonine	Thr
Glycine	Gly	Tryptophan	Trp
Histidine	His	Tyrosine	Tyr
Isoleucine	Ile	Valine	Val

The primary structure of proteins is determined by the sequence of amino acids. For example, the hemoglobin protein carries oxygen in the red blood cells and is actually made up of four protein chains. Below are examples of both the correct and incorrect sequence of amino acids in one of the protein chains of hemoglobin.

Correct
Sequence: val-his-leu-thr-pro-G L U-glu-lys...

Incorrect
sequence: val-his-leu-thr-pro-V A L-glu-lys...

The sequence or order of the amino acid is very important in the protein. The wrong sequence of one amino acids causes the protein to be different and non-functional. You can see that the *valine* amino acid is the incorrect amino acid. When this happens in the human being, the hemoglobin protein doesn't function correctly. The result is the person has sickle cell anemia and is unable to carry the correct amount of oxygen in the body.

Carbohydrates include the sugars, starches, and fiber. They are used for energy in plants and animals and also as structural components. Carbohydrates occur in three different forms. These are monosaccharides, disaccharides, or polysaccharides.

Monosaccharides are   simple sugars that contain one   ring-like structure of carbon atoms with oxygen and hydrogen atoms attached to the carbon atoms. The

27

three most common monosaccharides are: (1) Glucose also called corn sugar or dextrose. (2) Fructose from fruits. (3) Galactose from milk.

Disaccharides are two simple sugars chemically bonded together. Three examples are: (1) Sucrose from plants made up of glucose and fructose. (2) Lactose from milk made up of glucose and galactose. (3) Maltose from grain seeds made up of glucose and glucose.

Polysaccharides are long chains of monosaccharides (simple sugars) chemically bonded together. Some examples are: (1) Starch is a storage form of glucose found only in plants. A starch molecule contains 3,000 to 18,000 glucose molecules. So a starch molecule is very large. (2) Glycogen is the storage form of glucose found only in animals. (3) Cellulose is structural component found in plants. Animals can't digest it.

Lipids are water insoluble substances. We will look at just four common groups of lipids.

Fats and oils (or triglycerides). Fats are solids at room temperature and come only from animals. Some examples are butter, margarine and triglycerides which are found in your blood. Oils are liquids at room temperature and come only from plants. Some examples are corn and olive oil.

Phospholipids are another group of lipids with a similar structure to fats and oils except that they contain the phosphate group (composed of a phosphorus atom surrounded by four oxygen atoms). Phospholipids are important in living things because they make up the cell membranes.

Steroids contain four rings made up of carbon. Examples of steroids are cholesterol (precursor of other steroids), testosterone (the male reproductive hormone), and estrogen (the female reproductive hormone).

Nucleic Acids are information molecules found in cells of all living things. Deoxyribonucleic acid (DNA) and ribonucleic acid (RNA) are the names of the nucleic acids found in plants and animals. Chromosomes or genes are composed mainly of DNA. DNA and RNA contain the code for amino acids in a particular order and thus tell the body what kinds of protein to make. A nucleic acid is simply a chain of nucleotides strung together and are named for the base part of the molecule. There are only four different nucleotides in DNA named for the following four bases:

Cytosine	(abbreviated C)
Adenine	(abbreviated A)
Guanine	(abbreviated G)
Thymine	(abbreviated T)

An example of a piece of DNA molecule could be:

AAACTTGAGTCAT

These nucleotides are chemically bonded together. The most interesting thing about the DNA molecule is that it is the code for proteins. Each set of three nucleotides codes for an amino acid. In the above example AAA codes for lysine, an amino acid. The next set of three nucleotides or *triplet* - CTT codes for another amino acid and so on. All these amino acids are strung together (bonded chemically) to form a particular protein. The part of the DNA molecule that contains a group of nucleotides that codes for a protein is called a *gene.* Humans have thousands of genes and each code for a different protein.

Conclusion

The complexity of chemistry goes beyond our brief introduction here. The physical properties of the individual elements, their roles in reactions, and the hundreds of compounds are all subjects of interest. Geologists, medical doctors, pharmacists, engineers, homemakers, and farmers all depend upon knowing something of the discipline of chemistry. Fortunately, chemistry is the specialty of many researchers in our world and there are millions of articles and books which represent the results of good scientific research. Our focus in this chapter was to provide the reader with the basic building blocks for understanding the creation and evolution models.

In the next chapter, we will use some of these building blocks to learn some more foundational material in the biological sciences.

Discussion Questions

1. How important is chemistry to understanding the arguments for creation and evolution models?

2. How do you explain the organization and complexity of chemistry apart from a Creator? Does it appear to be designed or a result of chance events?

Chapter 4

The Principles of Biology

Biology is the study of life. As we discussed in Chapter 2, there are several areas of study within biology. For example, an area can be concerned with a particular type of organism (e.g., herpetology is the study of reptiles), or it can be concerned with a relationship among organisms and the physical world (e.g., ecology).

All of these areas of study are very complex and literally hundreds of thousands of articles and books can be found that describe and explain these areas. However, all these publications are dealing with one or more of the basic concepts of biology: (1) the organization of life, (2) the classification of the organisms, (3) structure and function of the organism, or (4) the mechanism of inheritance. Many, if not most, of the arguments for evolution and creation center around conclusions drawn from the principles developed in these areas.

Therefore, this chapter will provide the basics of biology in order for the reader to better understand these arguments.

The following characteristics define all living things:

- All life contains DNA
- Cells are the basic unit of all life
- Living things are adaptable to a changing environment

Living things are found in the earth's *biosphere*. The biosphere includes both the earth's crust and the sky. *Organisms*, whether plants or animals, can be studied as individuals or in a group. A group of the same kind of plant or animal is referred to as a *population*. Organisms are usually found together, coexisting with other organisms in a small area, such as a pond. This is referred to as a *community*. The community of organisms, along with their environment (climate, geology, etc.) form an *ecosystem*. Large areas of similar ecosystems are called *biomes*. Examples of biomes are deserts, deciduous forests, or tropical rain forests.

Classification of Living Things

Scientists organize and name living things in an orderly manner. *Taxonomy* is the process of classifying organisms in established categories. Biologists use the binomial system of classification which was invented and published by Carl von Linne (also known as Linnaeus) in 1735.

The kingdom category is the highest division of living things and subsequent categories get progressively subordinate.

There are five kingdoms. Within each of these kingdoms, every living thing can be classified:

Prokaryotes (monera) are simple (no nucleus), one celled bacteria and cyanobacteria (blue green algae). They have cell walls.

Protista are one celled, photosynthetic or heterotrophic (eat other organisms) such as protozoa, amoeba, all algae, diatoms, slime and water molds.

Fungi contain filaments and cell walls. They include the molds, mushrooms, yeasts.

Animalia are multicellular and have cell membranes instead of cell walls. Some examples range from insects to mammals.

Plants are multi-cellular, contain a cell wall and chlorophyll for photosynthesis. Some examples are mosses, ferns, trees, flowers, and grasses.

Viruses can be placed in a separate kingdom or in with the prokaryotes. Viruses are not composed of cells, but only DNA or RNA, and a protein coat.

The kingdoms are subdivided into *phyla*. Within the phyla, there are *classes* of organisms. Each class contains various *orders* and within the orders there is another subdivision called *family*. It is within the families that the basic units of classification are found -- the *genus* and the *species*. A genus is a taxonomic grouping of many species that are similar. There are two specific names, a genus name and a species name that are used to classify every organism, plant or animal. This name is unique and not given to another plant or animal. The species is a group of animals or plants that are very similar, almost identical. They usually look alike and have very similar genes. By definition, they can interbreed and reproduce offspring with one another but not with different species. For example, dogs can interbreed with other varieties of dogs. However, dogs can't interbreed with cats.

The following table illustrates the classification system as it applies to three examples:

Taxonomic Classification	Human	Sweetcorn	Monarch Butterfly
Kingdom	Animalia	Plantae	Animalia
Phylum/division	Chordata	Anthophyta	Arthropoda
Class	Mammalia	Monocotyledons	Insecta
Order	Primate	Commelinales	Lepidoptera
Family	Hominidae	Poaceae	Danaidae
Genus	*Homo*	*Zea*	*Danaus*
Species	*sapiens*	*mays*	*plexippus*

Further subdivision within orders provides a refined classification. Humans are differentiated from monkeys and apes in the following way:

Within the order Primates, there are two suborders:

> (1) Prosimii - lemurs and tree shrews
> (2) Anthropoidea - monkeys, apes and humans

Within the suborder Anthropoidea, there are three superfamilies:

> (1) Ceboidea - new world monkeys
> (squirrel monkeys and spider monkeys)
> (2) Cercopithecoidea - old world monkeys
> (baboons and rhesus monkeys)
> (3) Hominoidea - apes and humans

Within the superfamily Hominoidea, there are three families:

> (1) Hylobalidae - gibbons
> (2) Pongidae - chimpanzees, gorillas, and orangutans
> (3) Hominidae - humans

The classification system is a good way to categorize organisms for the purposes of study. Humans have many characteristics in common with monkeys and apes; therefore, we are classified as members of the same taxonomic superfamily. Remember, the taxonomic system is man-made and shouldn't be used to make inferences about ancestry. The existence of a species (living or extinct) is not dependent upon evolution and is certainly not dependent upon a system which implies a relationship of one organism to another. Taxonomy does not and will never be a system for inferring or conferring relationships between species. Unfortunately, as we will see in Chapter 6, the evolutionists have contrived lineage and classifications which attempt to do just that.

Structure of Living Things

Complex organisms, regardless of their classification, are made of the same basic four-level structural hierarchy:

> *Organ system* such as the circulatory system (heart, lungs, blood vessels).
>
> *Organs* such as the lung , the heart, the kidney, etc.
>
> *Tissues* such as muscle, nervous, connective, and epithelial
>
> *Cells* such as muscle cells, bone cells

Within the cells there are *cell organelles* such as mitochondria and the cell nucleus. Certain *molecules* such as glycogen, glucose, and fat are also found inside the cell.

Cells and Their Composition

Cells are the basic unit or building block of all living things except viruses. Cells are really where all the action takes place. Cells are where the food is burned and energy is released for use. Most disease processes occur at the cellular level. When evolutionary biologists talk about evolution of animals and man, the evolution they are talking about is thought by them to be occurring at the cellular level. We need to know what is going on in cells to understand the complexity of life. The following are descriptions of some of the major cell organelles which comprise cells and serve various functions. An electron micrograph picture of the inside of cell is shown in Figure 5. The majority of the picture is of the nucleus.

The *cell membrane* surrounds and holds the cell together. It is selectively permeable; that is, it allows only some substances to pass through. The membrane structure is made up of a double layer of phospholipid (fat) molecules.

The *endoplasmic reticulum (ER)* is a complex network of membranes for transport of molecules in the cell. The rough endoplasmic reticulum has ribosomes on it. The smooth endoplasmic reticulum is where lipid synthesis occurs.

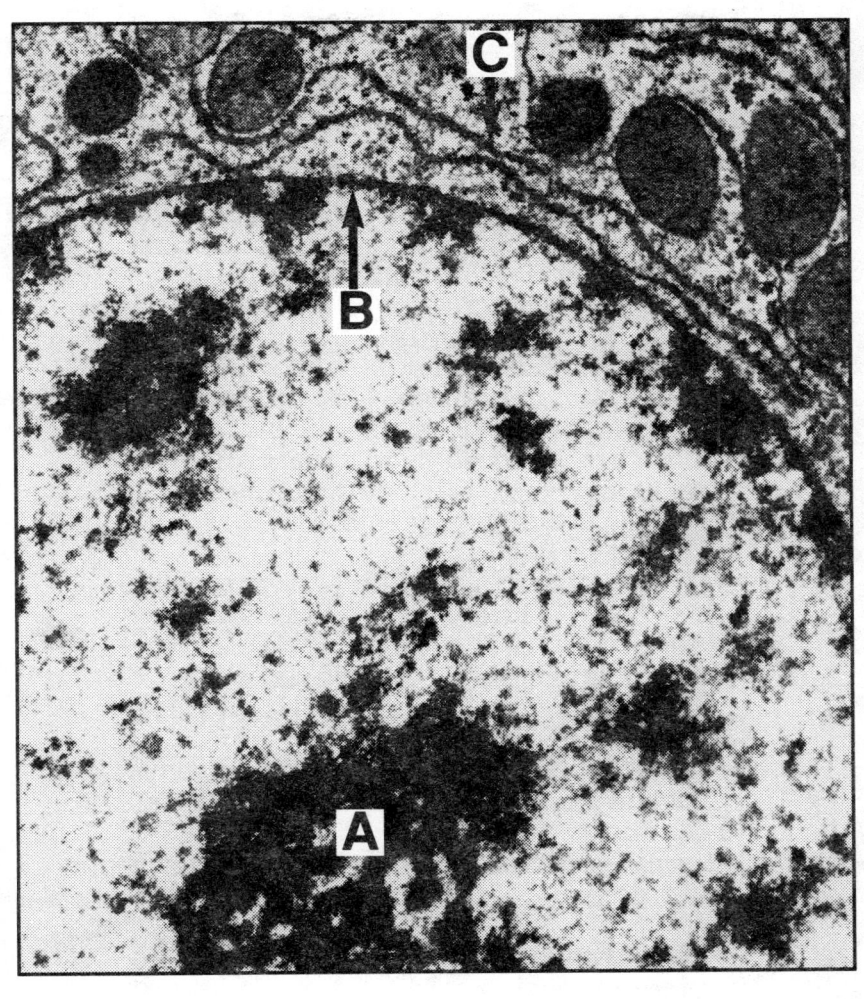

A - Chromatin (DNA); B - Nuclear Membrane;
C - Endoplasmic Reticulum

Figure 5.   Electron Micrograph of a
Cell.

Ribosomes are composed of protein and nucleic acids. They are the protein factories of the cell. The ribosome manufactures proteins from the components of amino acids. The ribosomes are located in the endoplasmic reticulum and in the cytoplasm (the substance between the cell membrane and the nuclear membrane).

The *Golgi apparatus* is found near the nucleus and is composed of a stack of about six membranous sacs. It packages the proteins made in the ER by adding on sugars. It then sends the protein to the exterior of the cell wrapped in a portion of the golgi membrane.

Mitochondrion are the powerplants of the cell. Energy in the form of Adenosine Tri-Phosphate is made here by oxidizing (burning) glucose.

Lysosomes are sacs of powerful enzymes that digest anything with which it comes in contact. It is the cell recycling center. For example, it decreases body tissues such as the uterus after delivery of a baby.

A *centrosome* is composed of two centrioles which are hollow cylinders made up of protein. They are involved in cell division and help chromosomes divide.

The *cell nucleus* is found in the center of the cell and contains the genes of the cells. It is like the brain of the cell since it directs the cell's activities. It contains the nuclear membrane, nucleolus and chromatin.

The *nuclear membrane* is a semipermeable double membrane, like the cell membrane, and it prevents the genes or DNA strands from leaving the nucleus and keeps other molecules out. The *nucleolus* is a small sphere of RNA and protein. It synthesizes ribosomes. The ribosomes then leave the nucleus through pores in the nucleus. There can be more than one nucleolus in the nucleus.

Chromatin is the DNA in the uncoiled state like a bunch of spaghetti or a mass of yarn. DNA in the coiled state is called a *chromosome*. It contains information (genes) on how to make a protein. There are 46 separate chromosomes in each nucleus of every cell in the human body (except the reproductive cells).

Cell Division

Chromosomes are composed of DNA in tightly wound strands. Each cell in an animal or plant contains the same number of chromosomes except for the gametes (eggs and sperm). The following table gives some examples of chromosome number:

Humans	46 chromosomes	(23 pairs)
Chicken	78 chromosomes	(39 pairs)
Goldfish	94 chromosomes	(47 pairs)
Corn	20 chromosomes	(10 pairs)
Cat	38 chromosomes	(19 pairs)
Mosquito	6 chromosomes	(3 pairs)
Dog	78 chromosomes	(39 pairs)

Homologous pairs of chromosomes are chromosomes that are the same size and carry identical functional genes such as the hair color gene. For example, the eye color gene can be for blue color on both chromosomes or blue color on one chromosome and brown on the other chromosome. These varieties in genes are called *alleles*. In other words brown, blue, and green are examples of the eye color allele.

For humans there are 46 chromosomes or 23 pairs. Chromosomes have a number assigned to them; 1 through 23. The number 23 chromosomes are special and are called *sex chromosomes*. They are referred to as the X and Y chromosome. All 23 pairs of chromosomes are found in every cell in the body, but the genes of all the chromosomes are not in use in every cell. For example the eye color gene is not in use in the liver cells but only in the eye cells. An exception to the chromosome number occurs in gamete cells (See Meiosis below).

Mitosis is the process of nuclear and chromosome division. One original cell divides into two new cells which are genetically identical. In humans the original cell has 46 chromosomes (23 pairs) and the two new cells have 46 chromosomes (23 pairs) each.

Meiosis is the process by which gametes (sperm and egg) are formed. The chromosome number is halved in animals. For example, in humans 46 chromosomes are reduced to 23. This occurs so that when the egg cell and sperm cell combine at fertilization, there is a total of 46 chromosomes. Meiosis only occurs in the ovary of the female and in the testes of the male.

Genetics

Genetics is another subset or specialty of biology. It is the study of how characteristics or traits, such as hair color, are passed to the offspring of living things. A trait is a certain characteristic in any living organism such as hair color or it can be a particular protein in the body such as insulin. *Traits* are controlled by a gene on a chromosome and traits are inherited.

Examples of inherited traits:

- Eye color
- Blood proteins such as hemoglobin
- Skin color
- Blood type
- Instincts such as thirst or hunger

A *gene* is a region on the chromosome comprised of a series of nucleotides. Most genes code for proteins. It is the protein that is responsible for the trait. For example, the eye color gene codes for a protein that causes the eye to have a certain color. When studying genetics, the gene can be symbolized by a letter of the alphabet. For example, a pea seed can have the gene which makes it a yellow seed or a green seed. A capital Y is used to represent the yellow seed because it is the dominant seed color and a small y is used to represent a green seed, since it is recessive.

Y = yellow seed   y = green seed

An *allele* is an alternative form of the same gene. In this example, the yellow color allele or the green color allele in the pea plant (Y or y). It's possible for organisms to possess both alternatives, even though it only exhibits one of them. A *genotype* is the actual genes that the plant or animal cell contains. There are always two alleles because the chromosomes that are comprised of these genes are paired. One allele (gene) is on one of the chromosomes and the

44

other allele is on the other. For example, a yellow pea seed could be YY (both alleles for yellow) or Yy (one yellow and one green). However, since Y is dominant, the plant will have yellow seeds. On the other hand, the *phenotype* is the expression of genes in the animal or plant. For example in the pea plant, the phenotype would be a yellow pea seed or a green pea seed. It is the trait of the plant or animal that is expressed or seen in the plant or animal.

Mendelian Genetics

Mendelian genetics is fairly simple and straightforward genetics that was first discovered and explained by Gregor Mendel. Gregor Mendel was a Catholic priest in Czechoslovakia in 1856-1864. He was hired to breed fruit plants such as peas because he was also a biologist and mathematician. He decided to study pea plants in detail. He knew something controlled traits in pea plants and he called them factors which we call genes today. He found each trait was controlled by two factors which we know are the two alleles of the gene. He used math to analyze his results of his breeding of many pea plants. He discovered that one allele was dominant and one was recessive for each of seven traits that he studied in the pea plant.

Examples of traits that Mendel studied include:

T = tall pea plant   t = short pea plant

T t are tall plants called heterozygous since both genes are present.

T T are also tall plants called homozygous dominant since only the dominant genes are present.

t t are short plants called homozygous recessive since only the recessive genes are present.

Y = yellow seed   y = green seed

Yy have yellow seed called heterozygous

YY have yellow seeds called homozygous dominant.

yy have green seeds called homozygous recessive.

Modern Genetics

There are some exceptions discovered since Mendel. Although not contradictions, modern genetics studies traits that do not follow the simple principles of Mendelian genetics. There are several illustrations of modern genetic principles, including *polygenic inheritance*, *multiple alleles*, and *co-dominance*.

Polygenic Inheritance. This is where two or more genes control a trait. Some examples are skin color, eye color, and height. More than one gene controls height and, as a result, there are not just two different heights in humans which is what there would be if height was controlled by one gene. There is a wide range of heights which tells biologists that there is more than one gene for height. It is known that tall genes are recessive to short genes for height. Eye color is determined by the amount of melanin in the eye. The more genes one has that have the allele for melanin, the darker the eye color which ranges from dark brown (most melanin) to light brown to hazel to green to blue (which hardly contains any melanin). Red eyes do not contain any melanin at all. Skin color is a similar situation. Light skin contains hardly any melanin because there are few genes coding for melanin in those cells. Dark skin has more genes that code for melanin.

Multiple Alleles. This is where more than two alleles are possible for a trait. For example:

> Blood type A allele produces A protein
> Blood type B allele produces B protein
> O allele produces no protein.

A and B are both dominant alleles and O is recessive. Therefore, there are many blood types or blood phenotypes possible. Type O blood can only be OO genotype since O allele is recessive. Type AB blood is AB genotype because A and B allele are co-dominant. Type A blood can be AA or AO genotype. Type B blood can be BB or BO genotype.

Co-dominance. This is where there is no dominant or recessive allele. An example is the *four o'clock flower* with only two alleles:

R = red   r = white

The phenotypes are:

RR = red

Rr = pink

rr = white.

Conclusion

After seeing the beauty, complexity, and high degree of organization in the biological world, it's difficult to believe that anyone would try to explain life as a product of random chance events. However, that's exactly what the evolutionist does. It seems much more logical and reasonable to see the hand of a designer or Creator. This concludes the study of the chapters that present some basics of science. In the next chapter, we will examine the creation model, as described in Genesis. With an understanding of the principles of biology, the creation model should be better understood and appreciated.

Discussion Questions

1. Are there any biological principles that stand out as reflecting God's personal involvement? Why?

2. Does biology seem random and chaotic or planned and orderly? Have you ever seen anything orderly which happened by chance?

Chapter 5

The Creation Model

Both the creation model and the evolution model are comprised of numerous submodels, complex arguments, and a fairly large array of points of contention. Collectively, numerous authors have had a great deal of success in developing the case for the creation model over the evolution model. (Notably, Denton's book *Evolution: A Theory in Crisis*, Gish's book *Evolution: Challenge of the Fossil Record*, Baker's book, *Bones of Contention*, Morris and Parker's book *Scientific Creation*, Whitcomb and Morris's book *The Genesis Flood*, and many others.)

As we said in the preface, this book is not designed to be a comprehensive review of the literature on the controversy, nor is it a dissertation on any of the issues. Rather it is a book that provides the reader with some elementary instruction that allows an understanding of the issues. Also, the intent is to present the basic premises of both models in the hope that the deficiencies in the evolution model and the logic of the creation model would become apparent.

There are two main areas that define both the creation model and the evolution model. These are:

- The Origin of Life
- The Explanation for Different Forms of Life

These are well established points of contention between creation and evolution. If there is one place the proponents of evolution and the proponents of creation agree, it's that these points are where the arguments center. (National Academy of Sciences, 1983; Scott, 1996; Morris & Parker, 1987; Baker, 1990; and others).

In this chapter, we will examine the evidence from the creation model's perspective. In fairness, these same points will be examined in Chapter 6 when we look at the evolution model.

The Bible Outlines the Premises for Creation

Understanding the creation model requires us to first examine the scriptures that describe the creation. Next, we must put this description into scientific terms. This will serve to define the creation model. From this model, there should be certain evidences, outside the scriptures, that support or contradict the model. Finally, we will see how the creation model explains the two points of contention.

The First Day of Creation (Genesis 1:1-5): Heavens, Earth, and Light

> 1   In the beginning God created the heavens and the earth. 2   Now the earth was formless and empty, darkness was over the surface of the deep, and the Spirit of God was hovering over the waters. 3   And God said, "Let there be light," and there was light. 4   God saw that the light was good, and he separated the light from the darkness. 5   God called the light "day," and the darkness he called "night." And there was evening, and there was morning --the first day. (NIV Genesis 1: 1-5)

From the first verses in the Bible, we learn that God created heavens, earth, and light. The word *heavens* can be interpreted in a number of ways. In this context, heavens probably refers to the whole universe that exists apart from the earth (Davis). The word used for heavens is the Hebrew word *shamayim* which means "the firmament, which appears like an arch spread out above the earth" and refers to the universe (Unger, p. 462).

Next, the earth was created but without form. The Hebrew word used for earth in this scripture is *erets*, which can be translated as the whole world (Unger). This is the world we dwell in as opposed to the sky or heavens (Davis, Tenney).

In these verses, the third object of creation is *light*. An interesting note here is that the sun, moon, and stars have not yet been created, yet there is light. One can only speculate, but it seems that light existed before the sun and moon were "assigned" the job of providing light (Genesis 1:14-19).

God formed the creation from His command. This creation principle from Genesis is represented by the writer of Hebrews:

> By faith we understand that the universe was formed at God's command, so that what is seen was not made out of what was visible (NIV Hebrews 11: 3).

There are other scriptures which support this concept, such as Job Chapter 38, and especially, Job 38:4 in which God speaks to Job about Creation:

> Where were you when I laid the earth's foundation? Tell me, if you understand. (NIV Job 38:4)

Second Day of Creation (Genesis1:6-8): Sky and Vapor Canopy

> 6 And God said, "Let there be an expanse between the waters to separate water from water." 7 So God made the expanse and separated the water under the expanse from the water above it. And it was so. 8 God called the expanse "sky." And there was evening, and there was morning--the second day. (NIV Genesis 1: 6-8)

Verses 6 and 7 both refer to an expanse which separated two waters--the waters above and the waters below. These waters were part of the matter creation in verses 1 and 2. The waters below the expanse seem easily understood as the earth's surface water, but what about the waters which were above the expanse? This is an important concept which is referred to as the vapor *canopy* (Baker, p. 32; Petersen, pp. 26-27; Whitcomb and Morris, pp 121 and 254-257). This might be thought of as water vapor or cloud layer above the earth. It served as a shield from the direct sunlight and the sun's harmful rays. This formed a virtual greenhouse effect on the earth resulting in warm uniform temperatures. There would be no rain and no polar cold conditions. The scriptures themselves support this concept in at least four places.

First, in Genesis 2: 5-6 referencing the Garden of Eden "and no shrub of the field had yet appeared on the earth, and no plant of the field had yet sprung up, for the LORD God had not sent rain on the earth and there was no man to work the ground, but streams came up from the earth and watered the whole surface of the ground." No rain fell, but instead streams (also translated mist) were the means of watering the earth at this time.

Second, after the Flood, God made the following statement in Genesis 8: 22, "While the earth remains, seedtime and harvest, and cold and heat, and summer and winter, and day and night shall not cease." This is a description of the four seasons which, prior to the fall of vapor canopy, would not have existed.

Third, it was apparent that no one had seen a rainbow prior to the Flood (which we shall discuss later). In Genesis 9:13 God made a covenant between Himself and the earth that never again will water become a flood to destroy all flesh. He said "I set my bow in the cloud." As we all know, a rainbow requires the sun to shine and cast its rays on the clouds. Apparently, the vapor canopy blocked the direct sun light and also, there was neither rain nor cloud.

Fourth, the vapor canopy collapsed at the flood of Noah creating some of the flood waters. Genesis 7:11 says "the floodgates of the sky were opened." Thus the vapor canopy ceased to exist after the flood and of course does not exist at present. This canopy of vapor explains more clearly phenomena about the natural world in pre-flood times. If the vapor canopy existed, but fell at the time of the Flood, then we should expect to see certain evidences of that and should be able to explain the following phenomena:

> *Long lifespan of humans in pre-flood times.* The vapor canopy was similar to our ozone layer in that it filtered out harmful radiation. This explains why people and probably animals lived longer. There would have been no harmful mutations caused by

radiation and therefore animals and man would not have cancer and many other disorders which we now know come from genetic mutations. The ages of the men reported in the pre-flood genealogies are in the 800s and 900s. For example, Seth lived to be 912 years old. Enosh died at 905. Kenan was 910 when he died. Mahalalel died at 895, but his son, Jared lived to be 962. Harmful radiation may cause us to grow old quickly at present because the genes we need for everyday functioning could be broken down by radiation.

Large plants and animals in pre-flood times. The greenhouse effect of the vapor canopy would allow for tremendous growth potential for organisms. Examples of large plants and animals found in the fossil record include:

> Moss 3 feet tall
> Ferns 50 to 70 feet tall
> Giant Club Moss 100 feet tall
> Dragonflies with a 27 inch wingspan
> Cockroaches 1 foot in diameter
> Rhinoceros 17 feet tall
> Dinosaur -- Very large!
> Storks 20 feet long
> Beaver 7 feet 6 inches long

The climate of the pre-flood times. If there were a vapor canopy covering the whole earth there would be no extremes in temperature on the earth. There would be a uniform warm temperature all over the earth from pole to pole. There would be no high and low pressures because temperature was the same. Differences in temperature cause high and low pressures and weather such as hurricanes, etc. There would have been no rain and no weather as we know it. The fossil record shows tropical plants and large animals from pole to pole. The best explanation for this is that the temperature was warm all over the world which is what the vapor canopy theory would predict. There are numerous examples of fossilized tropical vegetation in now frozen arctic areas (e.g. Northern Canada-Palm trees in basalt; New Siberian Islands-fruit trees 60 feet tall).

Figure 6 is a drawing representing the concept of the so-called vapor canopy.

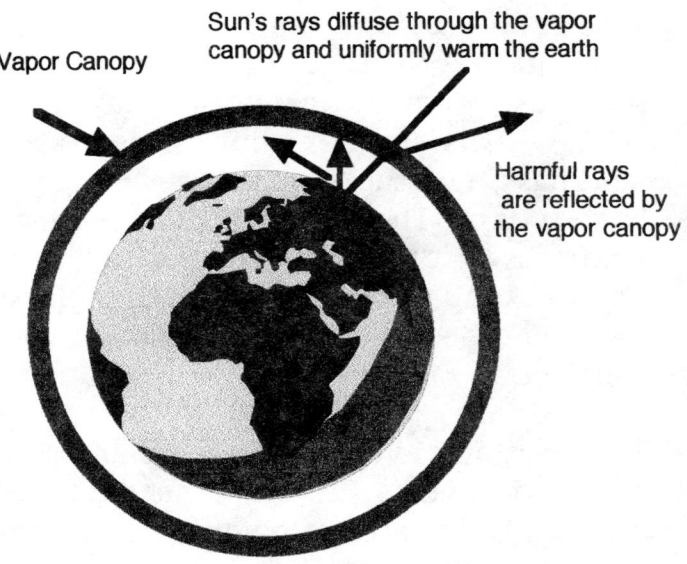

Figure 6.   Vapor Canopy Concept

Third Day of Creation (Genesis 1: 9-13)
Land, Seas, and Vegetation

> 9 And God said, "Let the water under the sky be gathered to one place, and let dry ground appear." And it was so. 10   God called the dry ground "land," and the gathered waters he called "seas." And God saw that it was good. 11   Then God said, "Let the land produce vegetation: seed-bearing plants and trees on the land that bear fruit with seed in it, according to their various kinds." And it was so. 12 The land produced vegetation: plants bearing seed according to their kinds and trees bearing fruit with seed in it according to their kinds. And God saw that it was good. 13   And there was evening, and there was morning --the third day. (NIV Genesis 1: 9-13)

We see that on the third day the land and seas were gathered into one place. These were previously created on the first day as a formless earth. This implies that there was only one landmass, not the separate seven continents we have today. There is geological evidence of one landmass in the past. The rock layers in South America and Africa match and their shape fits together. Geologists call this one large landmass Pangaea.

The creation of vegetation, the plants and trees, is recorded in verse 12. Scientists study this part of creation in the area of botany. Notice the phrases "according to their kinds." The plants and trees

58

made seeds and those seeds would grow another tree just like its parent, not a different one.

Fourth Day of Creation (Genesis 1: 14-19) Sun, Moon, Stars

> 14 And God said, " Let there be lights in the expanse of the sky to separate the day from the night, and let them serve as signs to mark seasons and days and years, 15 and let them be lights in the expanse of the sky to give light on the earth. " And it was so. 16 God made two great lights, the greater light to govern the day and the lesser light to govern the night. He also made the stars. 17 God set them in the expanse of the sky to give light on the earth, 18 to govern the day and the night, and to separate light from darkness. And God saw that it was good. 19 And there was evening, and there was morning - the fourth day.
> (NIV Genesis 1: 14-19)

On the fourth day God created the sun, moon and stars. We study this part of creation as the field of astronomy, which is the study of the planets and the stars. God first created the sun and the moon to give us light during the day and the night. Their other purposes were to create a daytime and a nighttime and to mark off the seasons and the years. God actually fixed a calendar for us. Our calendar is based on the revolution of the earth around the sun. Next God created the stars to give us light at night and to differentiate the night from the day.

Fifth Day of Creation (Genesis 1: 20-23) Marine life and Birds

20 And God said, " Let the water teem with living creatures, and let birds fly above the earth across the expanse of the sky. " 21 So God created the great creatures of the sea and every living and moving thing with which the water teems, according to their kinds, and every winged bird according to its kind. And God saw that it was good. 22 God blessed them and said," Be fruitful and increase in number and fill the water in the seas, and let the birds increase on the earth. " 23 And there was evening, and there was morning- the fifth day.
(NIV Genesis 1: 20-23)

Sixth Day of Creation (Genesis 1: 24-31) Animals, Snakes Reptiles, Insects, and Man

24 And God said," Let the land produce living creatures according to their kinds: livestock, creatures that move along the ground, and wild animals, each according to its kind." And it was so. 25 God made the wild animals according to their kinds, the livestock according to their kinds, and all the creatures that move along the ground according to their kinds. And God saw that it was good.

26 Then God said," Let us make man in our image, in our likeness, and let them rule over the fish of the sea and the

birds of the air, over the livestock, over all the earth, and over all the creatures that move along the ground. " 27 So God created man in his own image, in the image of God he created him; male and female he created them. 28 God blessed them and said to them, "Be fruitful and increase in number; fill the earth and subdue it. Rule over the fish of the sea and the birds of the air and over every living creature that lives on the ground."

29 Then God said, " I give you every seed-bearing plant on the face of the whole earth and every tree that has fruit with seed in it. They will be yours for food. 30 And to all the beasts of the earth and all the birds of the air and all the creatures that move on the ground-everything that has the breath of life in it- I give every green plant for food. " And it was so. 31God saw all that he had made, and it was very good. And there was evening, and there was morning- the sixth day. (NIV Genesis 1: 24-31)

Linnaeus was the first person who interpreted "kinds" to mean species in biology. He was born in Rashult, Sweden on May 23, 1707 and named Carl von Linne. He was the son of a pastor and had great respect for the Bible. He became a physician and studied botany. His goal was to name and organize into a nomenclature system all the original Genesis "kinds" in the Bible. He thought the species (kinds) could have variation in them such as different color varieties, but that the species were established and fixed. In other words he did not believe that the species could evolve into another different species. He is considered the father of biological taxonomy called the Linnean system which was described in Chapter 2. The Linnean system is the standard classification system of plants and animals.

On the fifth and sixth days of creation the Lord created animals, marine life , birds and man. This would include zoology (animals), and some subsets of zoology: mammalogy (mammals), ichthyology (fish), ornithology (birds), herpetology (reptiles), and entomology (insects). Notice that all the plants of the earth, not the animals, are given to Adam and Eve to eat. Plants were also to be the only food for animals.

What a perfect world this was with no bloodshed required to get food for either man or beast. Unfortunately, things did change and the perfect world became imperfect as we will see.

Fall of Man (Genesis 3: 17 -19)

17   To Adam he said, "Because you listened to your wife and ate from the tree about which I commanded you, `You must not eat of it,' "Cursed is the ground because of you; through painful toil you will eat of it all the days of your life. 18 It will produce thorns and thistles for you, and you will eat the plants of the field. 19   By the sweat of your brow you will eat your food until you return to the

ground, since from it you were taken; for dust you are and to dust you will return." (NIV Genesis 3: 17-19)

Man rebelled in this act of disobedience. This was an action that had far reaching consequences (see Appendix B) . We also begin to see changes in the creation. Man's diet is now restricted to the plants of the field which man must work very hard to grow and harvest. This is different from the beginning when God said man could eat any plant on the earth.

God said his creation was good many times in Genesis Chapter 1. Good means not evil, no problems, and perfect. We might interpret the good creation to mean there were no genetic defects in Adam or Eve. God would not create a genetically defective human. Therefore, there would be no genes for hemophilia, sickle cell anemia, cystic fibrosis, and the many other human disorders that we know are caused by defective genes. However, we see that due to the fall of man, the creation, along with Adam and Eve, are not perfect any more. In Romans 8:19-22 Paul speaks of the creation groaning and undergoing decay. Things only became worse.

The Flood (Genesis 6: 5 to 7: 24)

5   The LORD saw how great man's wickedness on the earth had become, and that every inclination of the

thoughts of his heart was only evil all the time. 6   The LORD was grieved that he had made man on the earth, and his heart was filled with pain.   7   So the LORD said, "I will wipe mankind, whom I have created, from the face of the earth --men and animals, and creatures that move along the ground, and birds of the air --for I am grieved that I have made them." 8   But Noah found favor in the eyes of the LORD. (NIV Genesis 6: 5-8)

11   In the six hundredth year of Noah's life, on the seventeenth day of the second month --on that day all the springs of the great deep burst forth, and the floodgates of the heavens were opened. 12   And rain fell on the earth forty days and forty nights.

23   Every living thing on the face of the earth was wiped out; men and animals and the creatures that move along the ground and the birds of the air were wiped from the earth. Only Noah was left, and those with him in the ark. (NIV Genesis 7: 11-12, 23)

From the time of Adam and Eve's rebellion, the earth became populated and things occurred which we can't explain with certainty. However, we know that things weren't what God wanted regarding the creation. Therefore, God judged man by natural forces and with a worldwide flood because of the wickedness of man. We do not know the exact geological events of the flood. However, there are many models, from a creation science viewpoint. These models are part of what is collectively referred to as the flood model or catastrophism. The geology of the earth today, (e.g., the Grand Canyon, lakes, and mountains) can be explained by the catastrophic event of the flood described in Genesis. The opposing model, which is subscribed to by evolutionists, is not based on the Bible. It is known as uniformitarianism or gradualism. This model proposes that the geology of the earth today is explained by millions of years of gradual erosion. We will discuss uniformitarianism a little more in the next chapter. The Biblical flood model does indicate the following events:

The vapor canopy fell. The vapor canopy described earlier was probably instrumental in providing the relatively "good" conditions of earth before the flood. As suggested by the scriptures in Genesis 7:11, the vapor canopy collapsed. There is no vapor canopy today. There certainly would be a lot of water from a fallen vapor canopy to help flood the earth quickly. Before the flood and the fall of the vapor canopy, radiation from the sun probably did not enter the earth because of the protective layer of the vapor canopy. Therefore, there was no harmful radiation at all for the humans and animals that lived on the earth. This may serve to explain the longevity of animals and humans discussed earlier. Genetic mutations would certainly shorten man's lifespan.

Initially, there was one large landmass and it was divided into continents after the flood. It's obvious today that we don't have one landmass, but rather divisions into what are called continents. It's possible that the catastrophic conditions of the flood could have had an effect on the separation. In Genesis 1: 9, we see the concept of one landmass presented:

Then God said,"Let the waters below the heavens be gathered into one place, and let the dry land appear" and it was so.

It is known in geology that there was one landmass in ancient times and geologists call it *Pangaea.*

Ice caps formed at the North and South pole. With the fall of the vapor canopy the weather on the earth changed dramatically. It is possible that ice deposition occurred rapidly and Ice caps formed--what geologist call the Ice Age.

The layers of the earth were formed by the flood. After the flood, water had to drain off and into the oceans so that land could once more be seen. It is possible that the displaced water (cubic miles of water) formed tidal waves and caused tremendous erosional forces. The receding water would have been powerful enough to have formed canyons such as the Grand Canyon and also river beds. Mt. St. Helens was a recent catastrophe here on earth and we can view it as an experiment. We know exactly what happened because scientists recorded the entire event both on camera and other scientific equipment. Many layers of earth were laid down in minutes and hours. A hundred years from now, someone looking at the layers formed could mistakenly assume that it took millions of years.

The Two Principles

At the beginning of the chapter, we said that both creation and evolution hinged on two points: (1) Origin of Life and (2) The explanation of different forms of life. Now that we've looked at the events of creation, let's examine some of the evidence that defines these two points from the creation perspective.

Origin of Life

In the six days of creation, which we just studied, we see that God created all living things. The Bible didn't say that God created some things and then let nature take its course. It clearly says that each was created according to its kind. Even man was created by God. This clearly defines the origin of life. There is no evidence to the contrary. The oldest civilizations excavated reveal man in the midst of all the creation and in the present modern form. There is immense agreement among scholars that a supernatural event (creation) took place. The alternative of evolution has numerous arguments against it, but we will have to wait to see the arguments in the next chapter.

Explanation of Different Forms (or Diversity) of Life

Once again, the scriptures are very clear that each species (kind) was created, not left to chance, for its form. The evolution model relies heavily on a phenomenon in nature called natural selection. This was a pillar of Darwinian evolution and served to further Darwin's position tremendously. Natural selection, as the name implies, is selection out of a population of plants or animals, the most fit for that environment. The interesting thing is that natural selection is, in fact, a real scientific phenomenon. While most scientists would discuss it as the mechanism of evolution, it stands on its own in the creation model as a means for survival of animals in different environments. Linneaus has noted that variations already occur within species. This is the basis for natural selection -- a naturally occurring variation of a characteristic or variety of alleles, genetically speaking. According to the creation model, the variety of alleles for each trait such as hair color were created.

A very common example is peppered moths in 19th century England. There exists two alleles for wing color-dark peppered and a light peppered color. Light peppered moths predominated in pre-industrial England. Since there was no black soot on the trees, light peppered moths blended in on the bark and were not eaten by birds, while the few dark moths were eaten more frequently. Then after the industrial revolution, the tree bark became black from soot and the light peppered moths were eaten, but dark moths survived and increased in numbers. Thus the number of black moths became greater than the number of light peppered moths. Nothing new was created, just a shift in the gene frequency of the dark and light peppered color alleles (more dark ones and fewer light ones). What is not presented in the textbooks is that if the conditions changed to favor the light colored moths, the ratio would change to more light ones and fewer dark.

Natural selection is in agreement with both the Bible and science, but natural selection doesn't give rise to new species. Natural selection merely allows a shift in the gene frequency in a population to favor certain traits which give its members a better survival opportunity. Sometimes the changes in the environment are extreme and place a tremendous amount of pressure on the population. When this pressure is so extreme, it may lead to a species extinction.

One can almost forgive Darwin of drawing the wrong conclusion about natural selection because, at the time he was working on his "theory," the mechanism for inheritance (through genetics) was being published by Gregor Mendel. Mendel published his work in a reputable science journal at about the time of Darwin and his theory of evolution. Mendel's work was read but ignored; the reason being that it conflicted with the theory of evolution which stated that new traits evolved. Later on in 1900 Mendel's paper was rediscovered and the evolutionary theory had to be rethought on how animals acquired new traits. The evolutionists reached a conclusion or idea that genetic mutations occurred to get a new trait. However, we know that this is not true, because although genetic mutations occur they are harmful or fatal in organisms. God created all the possible genes and in the pre-flood days there was much variety among the species of animals even dinosaurs. After the flood some animal species and varieties became extinct.

A final argument for the creation model's perspective on this is the fossil record. Numerous scientists throughout the ages have collected and studied the fossil record. There are fossils of organisms that are now extinct, however, this is to be expected given the changes that occurred after the flood. Also, in this century, we see hundreds of species becoming extinct. What the fossil record doesn't show us is any so-called transitional forms. There are no intermediates to suggest there was a transformation from one species to another -- not one! We will look at this argument again in the next chapter. One final thing shown in the fossil record is that there are examples of the very species we see today.

Discussion Questions

1. What do you think is the most convincing evidence that supports the creation model? Why?

2. What do you think is the most questionable part of the creation model? Why?

Chapter 6

The Evolution Model

The following definition of evolution is taken from a 1993 textbook of biology which was in wide use for introductory college biology: "the process that leads to structural and functional changes in species, making them better able to survive in their environment; also leads to the formation of new species" (Chiras, p. G-9). The concepts of evolution are generally attributed to Charles Darwin who only developed and popularized the ideas that many others already had. Still, Darwin, twenty years after serving as a volunteer naturalist on the *S. S. Beagle*, wrote and published his famous book *The Origin of Species*, in 1859. The first edition sold out (1250 copies) in the first day and the second edition also sold out quickly (1300 copies).

Much has been written describing the climate which would promote the sale of such a controversial work. Darwin himself may shed some light on the reasons. In his autobiography, Darwin suggests that he struggled with the idea that unbelievers in Christ would perish. He also said that the Old Testament "was no more to be trusted than the sacred books of the Hindoos

[sic], or beliefs of any barbarian" (From the Darwin Autobiography, deBeer, Ed., p. 49). He goes on to say that he felt driven to the conclusion that the old argument from design in nature "fails, now that the law of natural selection has been discovered" (From the Darwin Autobiography, deBeer, Ed.,p. 50).

Darwin made three observations which defined his theory of evolution: There is variation within a species of organisms; these variations can be passed on from one generation to another; and pressures to survive favored certain variations. This all seems so logical. As we discussed briefly in the previous chapter, the concept of natural selection is a sound one. Natural selection does occur; however, remember what Darwin didn't know was that at the same time he formulated the theory (1838-1859), Gregor Mendel was exploring the mechanisms for inheritance (i.e., genetics).

This lack of understanding of the principles of genetics kept Darwin in the dark with regard to the mechanism in which the inheritance took place. This explains the Darwinian concept that random mutation was responsible for the variations in organisms and these variations were either good or bad depending upon the natural selective conditions of the environment -- thus survival of the fittest.

In fairness to the evolution model, we will examine it on the basis of the same two points that we used in the creation model. We will view these points from the perspective of the evolution model. Likewise we will begin by defining the premises of the evolution model.

Overview of Evolution

The first principle is that nonliving material goes to living material. This is a basic premise since the idea of a Creator (God) is dismissed and instead, the very first life was generated by some set of circumstances. Incidentally, this concept is a very old one as well and was known for many years as spontaneous generation. There was an old theory that flies arose from dead flesh. Nobody saw the flies lay the eggs, only the emerging larvae and flies. Also, rats were thought to arise from garbage, etc. This was popular at the time of Darwin. Why not? If the origin of life was spontaneous, then why not flies from garbage? Darwin is quoted from his book saying:

> It is often said that all the conditions for the first production of a living organism are now present which could ever have been present. But if (and oh! what a big if!) we could conceive in some warm little pond with all sorts of ammonia and phosphoric salts, light, heat, electricity present that a protein was formed ready to undergo still more complex changes at the present day

such matter would be instantly devoured or absorbed which would not have been the case before living creatures were formed. (Quoted in Denton, p. 51)

The second principle is that diversity of life is explained by natural processes such as natural selection and mutations. Evolution says mutations improve life and cause it to evolve to more complex organisms. Evolutionists generally agree that it takes a long time for evolution to occur. This is the main reason for the claims of the earth being so old (perhaps 4.5 - 5 billion years or so). Darwin also indicated in his book that vast amounts of time would favor the events of evolution that he proposed. However, there is much evidence for a young earth. The dating methods used to prove the old age of the earth are questionable and we will look at some of these. Interestingly, the concept of natural selection is in agreement with both the Bible and science, but natural selection doesn't give rise to new species. As we pointed out in the last chapter, natural selection merely allows a shift in the gene frequency in a population to favor certain traits which give its members a better survival opportunity. The bottom line is that there are no transitional forms. Later in the chapter, we will look at the problems with some of the claims of transitional forms.

Nonliving Material Goes to Living Material

Scientists wanted to know how organic compounds and DNA were formed in order to advance hypotheses on how the first cell evolved. A. I. Oparin, a Russian scientist in 1924, provided evolutionists with an hypothesis called chemical evolution that explained how the first life was formed from organic chemicals. The theory stated that inorganic molecules in gaseous form would have reacted in the earth's primitive atmosphere to form organic molecules such as amino acids. The organic molecules then would have reacted with each other to form polymers. For example, amino acids would be formed first, then react with one another to form a polymer of amino acids which is called a protein. Likewise, simple sugar molecules would form a polymer known as a carbohydrate. Also, nucleotides would have reacted with each other to form the polymer -nucleic acid (DNA or RNA). From this point, the polymers would have self-assembled to form a complete living reproducing cell.

In the early 1950s, Stanley Miller, a graduate student of chemistry at the University of Chicago, decided to test Oparin's hypothesis. He set up a glass apparatus where gases carried by steam are sparked by electricity to simulate the early earth atmosphere. The gases used, which are simple raw materials thought to occur in earth's primitive atmosphere, are hydrogen, methane, ammonia, and water vapor. After a few days there was a brown liquid which contained many organic compounds including amino acids, urea, and lactic acid. Other experiments found that nucleotides could also be formed in the lab from inorganic compounds. Scientists have been able to form the polymers (protein, nucleic acids, and carbohydrates). Therefore, it appears that Oparin's hypothesis could have occurred on the earth. However the DNA formed does not contain any genes that are used in living cells. According to Morris and Parker, Miller's experiment not only produced some building blocks, but it yielded destructive amino acids in greater quantities which would have destroyed any good life-producing amino acids.

Scientists still cannot explain how polymers might self-assemble to form a living cell. Henry Morris developed a statistical argument that supports the improbability of life originating by chance. Even with generosity to the evolutionist in terms of the age of the universe (30 billion years old) and the number of particles available (10^{130} particles), his calculations and the contributions of several other studies conclude that the chance of life assembling itself by random chance events is zero.

This idea of creating life out of chemicals is very old and originated with the Greek scientist and philosopher Aristotle in the 4th century B. C. It is known as the theory of spontaneous generation. Everyone thought mice came from a pile of rags and flies came from garbage. This theory was disproved by Redi in the 1600s, Louis Pasteur in the 1800s and other scientists. It was replaced by the Law of Biogenesis which states that all living things come from other living things. So far the law still holds, no life has been created in the lab. Scientists today believe the Law of Biogenesis, but with one exception. They think that in the very beginning, new life (a cell) arose from organic polymers. If it occurred many years ago, then it should still be occurring at present. However, we see no evidence of this anywhere on earth. We do not see new cells arising from the oceans or ponds.

Diversity of Life Explained by Mutations

Evolution says mutations improve life, and cause transformations to more complex organisms. However, mutations are almost always harmful and cause life to degenerate. As we have said, natural selection is in agreement with both the Bible and science, but natural selection doesn't give rise to new species. There is no evidence in the fossil record for transitional species. Even if you give the point of an old earth, you would think, if evolution were true, that there might be one example among the millions of fossils that have been found. The truth is, there are none.

In Chapter 4, we explained the taxonomic classification system used to study and organize living things. By way of review, humans are classified in the following way:

Kingdom	Animalia
Phylum/division	Chordata
Class	Mammalia
Order	Primate
Suborder	Anthropoidea
Superfamily	Hominoidea
Family	Hominidae
Genus	*Homo*
Species	*sapiens*

The only member of the family hominidae is the human. The evolutionists suggest that there were other members of this family which are now extinct. One extinct group is classified as being members of the genus *Australopithicus* (translated - southern apemen). There is also a widely accepted view that there are several so-called extinct primitive forms of humans classified within the genus *Homo*. The modern humans (*Homo sapiens*) were thought to have had some predecessors, such as *Homo erectus*.

Furthermore, the theory held by many is that specifically, one species, *Australopithicus afarensis*, was the common ancestor of the humans and other extinct species of *Australopithicus*. These other *Australopithicus* species are classified *Australopithicus africanus* and *Australopithicus robustus*.

Australopithicus species are characterized as averaging 1.5 meters tall and having a brain size (cranial volume) of between 300 and 500 cubic centimeters. (Note: This is very ape-like. The human cranial capacity runs in the neighborhood of 1450 to 1500 cubic centimeters).

There have been a few claims, notably so-called missing links or transitional forms, that allegedly preceded modern humans--the so-called ape-men. Before we explain the problems associated with these, we can assure you that none of these claims are valid -- none!

Figure 7 shows the approximate locations where the fossils were found.

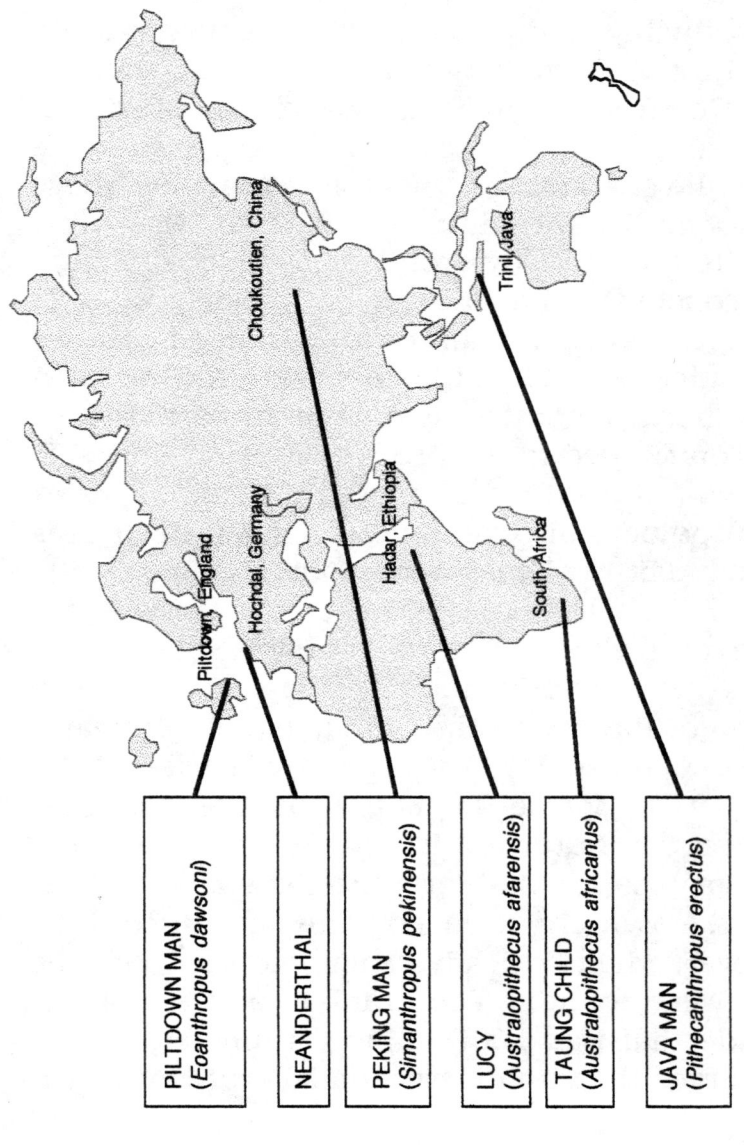

PILTDOWN MAN
(*Eoanthropus dawsoni*)

NEANDERTHAL

PEKING MAN
(*Simanthropus pekinensis*)

LUCY
(*Australopithecus afarensis*)

TAUNG CHILD
(*Australopithecus africanus*)

JAVA MAN
(*Pithecanthropus erectus*)

Piltdown, England

Hochdal, Germany

Choukoutien, China

Hadar, Ethiopia

South Africa

Trinil, Java

Figure 7. Approximate locations where some of the so-called "missing links" were found.

81

Nominations for Transitional Species for Humans

Neanderthal is an example of the first attempt at producing and exploiting a missing link or transitional form between the human and the ape. The first Neanderthal skeleton was found in the Neander Valley -- a limestone gorge with the Dussel River running through it -- near the village of Hochdal, Germany. This find was first exploited in 1857 by Professor D. Schoafhausen, who concluded that it must be human. Later in 1864, a German anatomist named Mayer concluded that Neanderthal suffered from a vitamin deficiency causing the disease rickets that contributed to the prominent brow. In 1872, Virchow, the father of modern pathology and anthropology, also presented a scientific paper arguing that the bones were modern.

Since that time, numerous other Neanderthal bones have been found in many countries on three continents (Iraq, China, Israel, Hungary, Greece, Central and North Africa). Between 1908 and 1913, Marcellin Boule of the National Museum of Natural History in Paris issued a scholarly paper that presented Neanderthal as ape-like which pleased the evolutionists. However, when critically examined, the following facts were found: The cranial capacity of the Neanderthal was 1600 cubic centimeters. This was larger than a modern human cranium which averages between 1450 and 1500 cubic centimeters. In 1957, the bones were reexamined by Strauss and Cave, anatomists at St Bartholomew's Hospital medical college (London). They determined that the bones, especially the foot were artificially positioned to give the appearance of being ape-like with a prehensile (opposable large toe). Also, the assessment of a diseased condition was also substantiated. It is now recognized by all scientists that Neanderthal was in fact fully human and not ape-like. Furthermore, there is evidence that Neanderthal had a complex society and religion. (Pitman)

Java Man (Pithecanthropus) was the next "nomination" for a transistional species. The existence of an ape-man was actually imagined by the German philosopher Haeckel in the 1880's and thought that it might be found in Southern Asia or maybe Africa. He even commissioned an artist to paint a picture of the imagined missing link. Haeckel's student in medical school, Eugene DuBois, went looking for it in the Dutch West Indies as a member of the Royal Dutch East Indies Army. After two years of searching in Sumatra, he was transferred by the Dutch army to Java. In the fall of 1891 on the bank of the Solo River near the village of Trinil, a molar tooth was found and a month later a fossil skull cap was found. Dubois originally thought these were chimp fossils. A year later a femur (thigh bone) was found 15 meters (about 50 feet) from where the skull cap was found. The analysis of the finds showed the femur to be from an upright walking modern human and the skull cap appeared to be ape-like. Thus he concluded that the two went together and was ape-man. He named it *Pithecanthropus erectus* (upright ape-man) and concluded that it was a precursor of humans. He exhibited the find in 1895. (Tattersall)

What he didn't say was that two fossilized human skulls were found at Wadjak (on the east side of the island) which would account for the human femur. He also found several more human femurs. After corresponding with Haeckel, he decided to forget about those human skulls and the additional human femurs. Virchow believed the skull was from a giant gibbon and that the femur had no connection with the skull. In 1921, Dubois admitted withholding information on the human skulls. Finally, in the early 1930s the human femurs were disclosed. In 1936 Dubois said that *Pithecanthropus* was actually a large gibbon. (Pitman)

Piltdown Man. The next attempt to provide the missing link was by Charles Dawson, a Sussex England lawyer and collector of fossils. For a couple of years, workers in a gravel pit near Piltdown, England would bring Dawson skull fragments which they had found. In 1908, Dawson claimed to have discovered human-like skull fragments. In 1911, he claimed to have found more. In 1912, Dawson gave his collection to Smith Woodard (Keeper of Geology at the British Museum). Woodard joined Dawson and a Jesuit paleontologist Pierre Teilhard de Chardin to conduct further work at the site. He found more bits of skull fragments and eventually found part of a right lower ape-like jaw bone minus the canine teeth. The jaw was broken in two places, at the hinge points and at the point. This made it difficult to associate with the skull fragments. Also, with the absence of canine teeth it couldn't be determined if it were ape or human (apes have large canines and humans have small ones). Still, Dawson pronounced it an ancestor of the modern humans and gave it the name *Eoanthropus dawsoni* (Dawson's Dawn Man). (Tattersall)

A couple of years later, in 1913, Teilhard, claimed to have found the missing lower canine tooth in the gravel pit and the three items were assembled as ape-man. Much debate ensued over the next several years; however, it wasn't until 1953 that it was found that the whole thing was a deliberate fraud. The bones were stained to give the impression of old age and the tooth was filed to present the ape-like shape. (Tattersall)

Peking Man has its beginning as a tooth purchased at a druggist's shop in Peking, China in 1903 by a Professor Schlosser. The tooth was discovered in a cave at Choukoutien, China. On this evidence, in 1927, a man named Davidson

Black secured a grant from Rockefeller, and established the Cenozoic Research Laboratory in Peking. Thousands of mammalian fossils and a few human teeth (575 boxes of bones) were collected through 1928. Eventually in 1929, a brain case was found to represent *Simanthropus pekinensis*, and even after Black's death in 1934, the team found more partial skulls (14 total) through the year 1937. (Pitman)

An interesting observation was made that many of the bones were charred (and found in a seven meter deep heap of ashes) and mixed with other mammal remains including deer. The skulls had signs of being struck. Marcellin Boule was invited to the site and his assessment was that the skulls were battered monkey skulls and were not more important than the other animal remains. He felt that this was evidence that modern humans had been eating the mammals and much of what what was found as fossils was from those meals. Also, the bones mysteriously disappeared and have never been seen since that time. (Pitman)

Nebraska Man also had its beginning as a tooth, found in America in 1922. On the basis of this tooth, it was determined, by the American Museum of Natural History, to be a mixture of human, chimpanzee, and *Pithecanthropus*. In 1925, when the trial occurred to determine whether evolution should be taught in schools, it was used to construct a picture of what the missing link would look like. It was named *Hesperopithicus*, however, later it was determined to be the tooth of an extinct pig. (Pitman)

Lucy and the *Australopithecenes.* The first australopithicene fossil was discovered and exploited in 1925 by anatomist Raymond Dart. A child's skull was found in a South African cave (called Taung) and thus the name Taung Child. It was given the scientific classification *Australopithicus africanus.* Subsequent to that discovery, many other similar fossils have been discovered, many of which were thought to be older. In 1974, at a site called Hadar in Ethiopia, Donald Johanson discovered a more complete set of bones classified as *Australopithicus afarensis*, (called Lucy). This was thought to be an older species and also was thought to have been bipedal (walked on two legs). This later point was criterion for being in the Hominid family. However, the brain capacity was very small and chimpanzee-like. Therefore, the conclusion was that this could be a missing link of sorts -- a transitional form that bridged the early apes with *Homo* -- early man. Lucy is one of the primary "missing links" being pushed today. Even among evolutionists there is a debate on the significance of these finds. There are several scientific criticisms of the find and the interpretation of the joints and bone structures.

One Nomination for Non-Primate Transitional Species

Considering the thousands of species alive today and the thousands of fossil species discovered, one would expect to find thousands of examples of transitional forms to bridge the so-called evolutionary gap from one species to another. This is the second main premise of evolution. As we pointed out, even Darwin fully expected to find these in abundance. The truth is that there are none. As we saw with the search for human "ancestors," the search has been futile. The search for a transitional form for all the other species is worse yet -- if that is possible. The only one that had (notice the past tense) a chance was the famous *Archaeopteryx.*

In 1860, the first of six fossils of *Archaeopteryx* was found in Solnhofen, Germany. For the evolutionist, it continues to represent a transition or intermediate between bird and reptile. Close examination revealed it had some skeletal features, such as a breast bone that was, in some ways, reptilian in form (*i.e.,* flat). However, many species of bird, such as ostriches, have similar features (*i.e.,* flat breast bones). *Archaeopteryx* also has teeth and three clawed fingers at the end of its wings; however, according to Charig (an ardent evolutionist) "the classic *Archaeopteryx* is surrounded by the impressions of unmistakable feathers and is therefore classified quite positively as a bird" (p. 83). He assesses *Archaeopteryx* as the oldest fossil bird and pronounces that it is NOT a reptile-bird intermediate.

Conclusion

If all of this seems confusing, it's because it is. Confusion starts when attempts are made to build a case for classifying extinct species into taxonomic categories based on fossil finds. Fragmentary evidence, uncertainty of the relationships among the bone fragments, and assumptions about the relative time periods when the fossils lived all lead to wild inferences. Furthermore, problems arise when inferences are made about the relationship of one fossil species and another contemporary fossil species. Finally, in the midst of all this uncertainty, the evolutionist suggests that there is "evidence" that these fossils can be "linked" as ancestors to modern humans.

Denton says "Without intermediate or transitional forms to bridge enormous gaps which separate existing species and groups of organisms, the concept of evolution could never be taken seriously as a scientific hypothesis" (p. 158). Even the staunch proponents of evolution acknowledged this from the very beginning. Yet no transitional forms have been found. This lack of transitional forms in the fossil record is undisputed among paleontologists. There has been a fervent attempt since Darwin's time to unearth a single fossil which demonstrates a transition from one species to another. There have been about 100,000 fossil species identified to date. These represent the fossilized remains of organisms that merely lived and died. As we will discuss in Chapter 7, fossils aren't exclusively extinct species, but are remains of anything that died and was put under conditions that allowed fossilization instead of decay or consumption by another organism. In fact, many fossils are classified as belonging to modern groups (phylum, class, family, etc.) Denton indicates that "It is still, as it was in Darwin's day, overwhelmingly true that the first representatives of all the major classes of organisms known to biologists are already highly characteristic of their class when they make their initial appearance in the fossil record" (p. 162). In other words, organisms haven't evolved.

Discussion Questions

1. List three reasons why the evolution model is weak.

2. What one thing seems to be a compelling argument in favor of evolution? Why is this argument weak?

Chapter 7

But what about ...?

In this final chapter we will briefly recap the conclusions of the previous chapters, but first we will address some so-called "loose ends." Whenever we have discussed this material with others or led studies on the subjects of this book, we often encountered interest in some subjects which didn't fit into any single chapter of the book. Among these: (1) The Age of the Earth, (2) Radiometric Dating Methods, (3) Fossils, (4) Dinosaurs, and recently (5) Life Originating on Mars.

The Age of the Earth

The suggested age of the earth began to get older at the same time that Darwin's theory of evolution was being proposed. The theory itself assumed a large amount of time would be available for the changes to take place. The long time needed for evolution was provided, in part, by a man named Charles Lyell in his book *Principles of Geology* which was published in 1830. In this book, Lyell proposed the theory of gradualism -- that the earth's features, such as canyons, river beds, mountains, and oceans can be explained by weather processes that occur today. These processes would take millions of years. Thus the rains, and cold and hot weather and snows would cause mountains to wear down and canyons to be cut out by moving water. But it would take a long time for this to happen. So Lyell reasoned that the earth would have to be millions of years old so that there would be enough time for the water to carve and cut out canyons and river beds (such as the Grand Canyon).

This concept is one of correlation. For example, let's assume you know that water runs from your bathtub spout at one gallon per minute (gradualism). If you enter the bathroom after the tub is filled, and measure the amount of water in the tub and find that it contains 10 gallons, you can easily calculate that the water ran for 10 minutes. Again, the assumption is that the water runs at one gallon per minute. However, your assumption could be wrong. Let's say that whoever was filling the tub was in a hurry and decided to add buckets of water from another source. The tub could have been filled a lot quicker. Likewise under Lyell's uniform conditions of wind and rainfall, the geologic formations of the earth can be explained as having taken a very long time. However, like filling the bathtub with buckets, a catastrophic flood could get the same results much quicker.

The reason that the age of the earth is so important is that even if evolution were possible, it would require a very long time. There is no dispute that nothing has evolved in the last 5,000 years. Evolutionists subscribe to an earth that is somewhere between 4.5 and 5 billion years old. However, there are a total of 68 estimates, based on observable scientific measures, that fall short of the required time for evolution. There are also at least 15 scientific measures which provide solid evidence of an earth that is less than 10,000 years old. For example, the rate at which nickel, silicon, and lead flow into the oceans coupled with the known quantities of these elements already in the ocean, provide evidence for an earth that is less than 10,000 years (Gish). In other words, if the earth were as old as evolutionists would like to think, there should be much more of these elements found in the oceans. Unfortunately, these measures are all based on the same assumptions of uniformitarianism or gradualism (i.e., the rates of decay or disintegration, or erosion are occurring at the same rate today as they were thousands of years ago).

The concept of gradualism also led to the concept that the deeper you go in sedimentary rock, the older the fossils are that were found in that strata (Figure 8).

Era	Period	Epoch	Start Time	Life Forms
CENOZOIC	QUARTERNARY	RECENT	10,000 yrs ago	
CENOZOIC	QUARTERNARY	PLEISTOCENE	2,500,000	Man
CENOZOIC	TERTIIRY	PLIOCENE	12,000,00	Grazing and Carnivorous Mammals
CENOZOIC	TERTIRY	MIOCENE	26,000,000	"
CENOZOIC	TERTIARY	OLIGOCENE	38,000,000	"
CENOZOIC	TERTIARY	EOCENE	54,000,000	"
CENOZOIC	TERTIARY	PALEOCENE	65,000,000	"
MESOZOIC	CRETACEOUS		136,000,000	Primates and Flowering Plants
MESOZOIC	JURASSIC		195,000,000	Birds
MESOZOIC	TRIASSIC		225,000,000	Dinosaurs and Mammals
PALEOZOIC	PERMIAN		280,000,000	
PALEOZOIC	CARBONIFEROUS PENNSYLVANIAN		320,000,000	Reptiles
PALEOZOIC	CARBONIFEROUS MISSISSIPPIAN		345,000,000	Fern Forests
PALEOZOIC	DEVONIAN		395,000,000	Amphibians and Insects
PALEOZOIC	SILURIAN		430,000,000	Vascular Land Plants
PALEOZOIC	ORDOVICIAN		500,000,000	Fish and Chordates
PALEOZOIC	CAMBRIAN		570,000,000	Shellfish and Trilobites
PRECAMBRIAN			4,650,000,000+	Algae, Eucaryotic Cells, and Procaryotic Cells

Figure 8. Geologic Time Scale. Adapted from Funk and Wagnalls New Encyclopedia Vol 11 , 1986

Gradualism is also the basis for explaining the evolution of organisms of lower complexity (found at lower levels) to organisms of higher complexity (found at higher levels). For example, trilobites (Figure 9) are found at the bottom of the time scale in the very early Cambrian period.

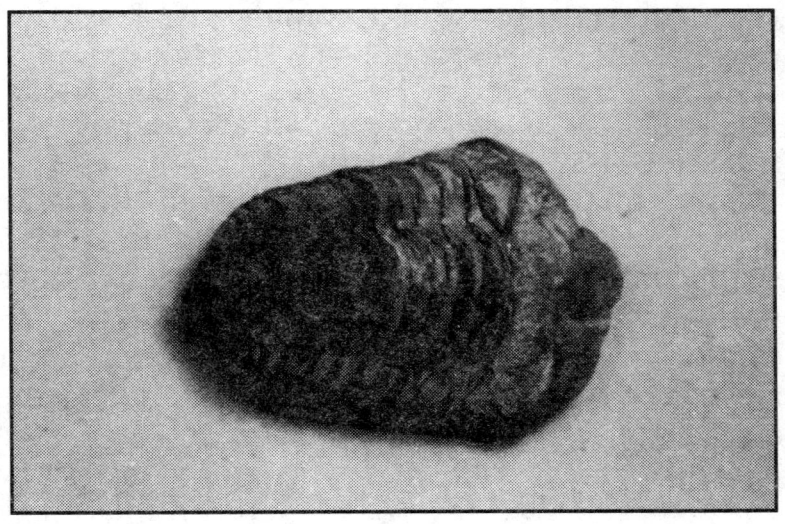

Figure 9.   A fossil trilobite
three   inches   long.

The evolutionist's explanation is that the trilobite was an early life form,thus found in lower strata. However, consider that trilobites were exclusively

95

ocean dwelling and lived on the ocean bottom. A better explanation might be that a catastrophic flood (i.e., the Genesis Flood), and its associated violent activities, would have silted over the lowest dwelling organisms first. The trilobites would have been prime candidates for a lower strata. If an organism's ecological niche and its capabilities for relocating (e.g., climbing a tree, or flying) are considered together, it makes sense that so-called higher organisms are found in higher strata. As we saw in the creation model, the various strata would contain the fossilized remains of life destroyed in the Flood.

The creation model presents a strong case for catastrophism. Gish and others point out that the age of the earth can never scientifically be proven. What can be said is that there is a significant variation in the results of the so-called geologic chronometers (or clocks). Therefore, the evolutionists can't be absolutely sure about their age of the earth. The 4.5 billion year age of the earth cannot be supported by any accurate means of measurement. This quote from the Physical Geology textbook sums up the weak arguments:

> The 4.5 billion year estimated age of the earth comes from various lines of evidence, mostly worked out by astronomers and planetary geologists who feel that all the planets formed about this long ago. (Plummer and McGeary, p. 154)

Without dwelling too long on this, we will point out some interesting frailties in this dogmatic statement by reprinting it with italicized emphasis placed on these areas of weakness.

> The 4.5 billion year *estimated age of the earth* comes from various lines of *evidence, mostly worked out* by astronomers and planetary geologists who *feel* that all the planets formed *about* this long ago.

This type of conclusion is not science and the old age of the earth cannot be supported.

Radiometric Dating Methods

There are two main radiometric methods we will discuss: (1) Uranium - Lead and (2) Carbon-14. Before discussing the mechanism of radiometric dating, let's review the definition of radioactivity. Atomic nuclei of radioactive elements, such as uranium 238 (92 protons and 146 neutrons), are unstable and break down into smaller atoms. For example, over time, uranium 238 loses 10 protons and 22 neutrons, leaving it with only 206 nuclear particles which is now nonradioactive lead (Pb - 206) which is referred to as a daughter product. The rate of so-called decay has been measured and is known for these radioactive elements. The time it takes for 50% of the element to decay completely (e.g., uranium to go to lead or carbon 14 to go to carbon 12) is called its halflife.

Uranium to Lead. This method is used to date rocks. Scientists assume that the halflife for U-238 is 4.5 billion years. This means that if you had 10 grams of U-238, after 4.5 billion years you would have only 5 grams. The main problem with radiometric dating methods is that in all cases it is unknown how much radioactive material was in the rock in the first place. For example, labs may have a starting assumption that there was 50% lead and 50% uranium (a 1:1 ratio) in the rock originally. Then, if there is 25% uranium in the rock at the present time then 4.5 billion years have elapsed since the rock was formed. However, there is no way of knowing the amount of the original percent of uranium or lead in the rock. Also, uranium and other radioactive elements can be lost from the rock by dissolving in weak acid washing over it through time. Scientists only know what the current rate of decay is and assume it has remained constant.

Carbon 14. Carbon 14 (C-14) which is radioactive is formed in the upper atmosphere from cosmic rays and nitrogen gas. Non-radioactive carbon 12 (C-12) is also present. Both of these carbons are found in carbon dioxide gas and are taken in by plants for growth. Animals eat plants and thus consume C-12 and also C-14. C-14 analysis actually dates the age of death of the organism; when the organism stops taking in C-14 and C-12. C-14 decays away and C-12 does not; so the method measures the ratios expected based on the decay rate. Less C-14 means an older specimen and more C-14 means a younger specimen. The halflife of C-14 is about 5,730 years. Since C-14 is made in the atmosphere, one assumption is that its availability has been constant over time, but that is not the case. It has been shown that volcanic activity causes variation in the amount of C-14 present. Also, in 1918 a comet fell over Siberia and the amount of C-14 in the atmosphere doubled. Solar flare activity and ozone pollution are two other factors shown to affect C-14 levels in the atmosphere. In pre-flood days when there was a vapor canopy that prevented entry of cosmic rays then there would have been little or no C -14. Preserved organic material from pre-flood times then would be dated as very old when actually they would be a few thousand years old.

Final Thoughts on Radiometric Dating

Radiometric dating methods such as uranium to lead can only date the age of formation of volcanic materials. Completely fossilized material can't be dated at all if all the organic material has become rock. Volcanic rock is dated; not the fossilized material. There are numerous problems associated with these methods. Sedimentary rocks (such as ones formed from a flood) can't generally be dated because of the possibility they could have been contaminated by previous existing or previously formed daughter products.

Other problems include the possibility of loss of material during weathering and heat.

Fossils

We talked some about the fossils in the creation chapter and the evolution chapter. As we said, the fossil record is a friend to the creation model. It explains very nicely that there was a catastrophic event in history (the flood of Genesis) which devastated life on the earth at that time.

Fossils are natural representations of organisms that have died. Any organism that dies is a candidate to become a fossil, however, many do not for a variety of reasons. One of the primary requirements to become a fossil is for the organism to be rapidly covered in order to prevent decay or consumption by other organisms. The sediment produced in floods is a good example of how this occurs. In fact, the flood of Genesis undoubtedly produced conditions which would entrap organisms in a way to promote fossilization. This knowledge is based on models available today from floods. The sediment produced would have covered organisms living at the time of the flood (of Genesis) and the fossilization process would then take place. It's not surprising that 99 percent of the world's fossils are found in sedimentary rocks formed by receding flood water.

Given the nature of the catastrophe of the Genesis flood, it's not surprising that there is an abundance of marine (or ocean) fossils found at high elevations and even in mountains. There must have been a tremendous upheaval during the flood event and we know that all the dry land was covered. When we were in Turkey, in the 1980s, we spent time climbing in the Taurus Mountains, north of Tarsus (the Apostle Paul's home town). Once, we were climbing at about 9,000 feet above sea level and found an abundance of fossil marine organisms including many types of shells and coral (Figures 10 & 11).

Figure 10. Fossil scallop shell approximately two inches in diameter; both halves intact.

Figure 11.   Fossil coral 3.5 inches long.

These were interesting discoveries for us, but there have been many such fossil finds in mountains around the world.

A similar discovery was made in the 16th century by Bernard Palissy who was a collector and student of fossil shells and fish. He wrote "I have drawn a number of pictures of the petrified shells that can be found by the thousands in the Ardennes Mountains, and not just shells, but fish....I have found more kinds of petrified fish, or the shells thereof, than I have of modern kinds now living in the sea" (Quoted in Abrams, p. 35). Something definitely happened to trap these large variety of organisms.

MacFall and Wollin present six main processes by which organisms become fossils. The following brief descriptions of these processes are provided:

Freezing is similar to putting food in the freezer. The decay process is suspended, and if frozen in ice, the organism becomes inaccessible to consumers.

Drying or dessication provides another aseptic (micro-organism free or clean) environment. This prevents the decay of the organism.

Original preservation is possible when the burial preserves the specimen both chemically and physically.

Petrification With this process, the original material of the organism is replaced by other material; or some other material fills the cracks and pores of the original organism.

Carbonization is a way that preserves soft-bodied animals and plants. The original composition of the organism is changed by bacteria, chemicals, heat, and pressure. What remains is a thin carbon film which provides a detailed representation of the original structure.

Casts and Molds are formed when an organism is trapped in sediment and in time it dissolves leaving an empty space or cast. After more time passes, this cast may refill with a replacement material and this forms a mold of the original organism's shape.

103

Figures 12, 13, 14, and 15 are other examples of common marine fossils.

Figure 12. Fossil fish 1.75 inches long from the Green River formation, Wyoming.

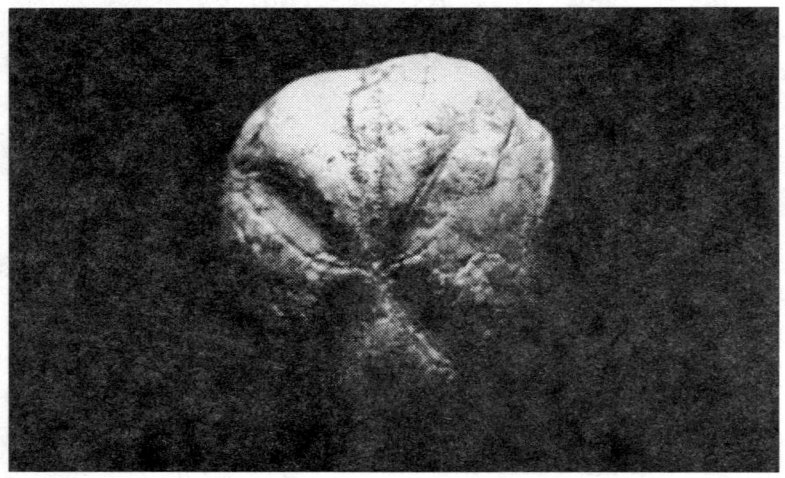

Figure 13. Fossil echinoderm (e.g. sand dollar, sea urchin) 2 inches in diameter.

Figures 14 and 15. Sedimentary rock samples from Waynesville, Ohio, showing a variety of common marine organisms.

105

Dinosaurs

The dinosaur was a member of the group of animals called vertebrates in the phylum Chordata. These animals had back bones composed of small bones known as vertebrae. Vertebrates also have a skull which encloses the brain.

Within the vertebrates there are many groupings of animals known as classes. For example, there is the mammal class, the fish class, the bird class, and the reptile class. The dinosaurs belong to the reptile class even though they are extinct. They were a larger reptile, more like a lizard than a snake or turtle. They were created along with the other animals and man, so they lived with man before the flood. If they were nonextinct before the flood, Noah certainly took representatives of them on the ark. Some may have lived for a time after the flood then became extinct. Evolutionary scientists place dinosaurs in the Triassic, Jurassic, and Cretaceous periods up to 235 million years ago (Refer to Figure 8).

For some reason, the dinosaurs disappeared (i.e., they became extinct). Scientists think the extinction might be due to the climate becoming colder and more unstable. Their reason for extinction fits very well into the creation/flood model. After the flood, the climate became colder and there was less vegetation on the earth for the herbivorous dinosaurs to feed on. They may not have adapted well to temperature change, had less food to eat, and thus died and became extinct.

Some general characteristics of dinosaurs are that they were land dwellers and did not crawl flat on the ground but were off the ground on two legs (bipedal) or four legs (quadrupedal). In most dinosaurs the forelimbs are much shorter and slighter than the larger hindlimbs.

A class of animals has subgroups called orders. There are many orders of reptiles including the order Chelonia (turtles) and order Squamata (snakes and lizards). Most of the orders of reptiles are extinct. Dinosaurs are classified as belonging to one of two orders, the Saurischia and the Ornithischia. The difference between the two orders is the way the hip bones are aligned.

The order Saurischia is divided into two groups, the theropods and the sauropods. The theropods were mostly carnivores and the sauropods were herbivores. Therapods were bipedal and contain the ferocious dinosaurs. Many theropods were not large like *Ornithomimus* the ostrich dinosaur at four meters in length. Larger ones included *Tyrannosaurus* (12 meters long, seven tons, and 15 cm teeth-the largest flesh eating animal to ever live), *Tarbosaurus* (12 meters long) and *Megalosaurus*. These carnivores ate other herbivorous dinosaurs. The sauropods are the typical dinosaur one brings to mind when a dinosaur is mentioned with a small head, long neck and very large body. They were most all quadrupedal herbivores. Famous ones are the *Brontosaurus* at 20 meters long and 30 tons, *Diplodocus* at 26 meters long and 10 tons and *Brachiosaurus* at 23 meters long and 80 tons.

The other order of dinosaurs the Ornithischia includes four groups. The ornithopods were herbivores and some were bipedal. They included the duck billed, parrot, and dome headed dinosaurs. The next three groups were quadrupedal, had armor for defense, and did not run fast. Firstly, were the horned dinosaurs such as *Triceratops* at nine meters in length and 5.4 tons. Secondly, were the plated dinosaurs which had thick plates that stuck out like spines all over their body; for example *Stegosaurus* at nine meters long. Thirdly, the armored dinosaurs with bony chunks and spikes on them such as the *Ankylosaurus* at 10 meters in length.

Life Originating on Mars

The story began in 1984 with a team of explorers in Antarctica who discovered and recovered a number of rocks which were subsequently identified as being fragments of a meteorite. This meteorite is presumed to have resulted from a collision between Mars and something else

16 million years ago. The resulting fragments then floated in space until falling to Earth 13,000 years ago. Among the rocks, one in particular, so-called Alan Hills (or ALH) 84001 has been the subject of some interesting claims. Twelve years after the discovery, NASA decided to announce in August 1996 that they believe that the meteorite showed signs that ancient life existed on Mars. Various lead stories went out in page one news such as "MARTIAN MESSAGE-THERE WAS ONCE ANCIENT LIFE HERE" (Cowen, Christian Science Monitor, Aug 8, 1996, p: 1, col: 1). Network and cable television were also running these stories. Here is the basis for these claims.

The NASA scientists leading the research, Dr. David McKay and Dr. Everett Gibson, largely base their arguments on the existence of two things found in the ALH 84001 rock. The first is what appeared to be polycyclic aromatic hydrocarbons (PAH) with traces of magnetite and iron sulfide with the PAHs. These chemicals can be associated with biological activity. On earth, PAHs are know to be formed by bacteria. The second evidence is what the researchers said were fossilized nanobacteria (i.e., very small bacteria) which were so small that an electron microscope was required to visualize them.

Some of the obvious questions arising from this include: How were time frames determined for the the events of the collision and the time in space? Also, how could the age of the rock be determined for a rock which was presumed to have been formed on another planet? At a meeting of the Planetary Division of the American Astronomical Society, the NASA team (Dr. McKay and Dr. Gibson) received what was called "blunt skepticism" from other scientists (New York Times, Oct 25, 1996, p. 24). When pressed on their evidence, several times Dr. McKay and Dr. Gibson admitted that they hadn't yet developed enough evidence to settle the controversies. This is certainly contradictory to an article released in the Atlanta Constitution which said "NASA scientists say a 'compelling' array of indirect evidence, culled from a Martian meteorite, makes a persuasive case that life can no longer be considered unique to Earth" (Aug 8, 1996, Sec A, p. 3). Subsequent articles refer to reports that the so-called microfossils weren't nanobacteria at all, but were magnetite crystals formed from high-temperature vapor processes (Wilson). Other articles report explanations arguing that the findings are similar to those found in volcanic vents and unlike those formed in biological activity (Cowen, 1997; Holmes).

Nevertheless, in December, 1996, the United States launched a NASA spacecraft to look for more clues on Mars. More expeditions are planned for Antarctica. Continued criticism appears in Science News (Feb 8, 1997) with contradictory evidence from scientists at both the Georgia

Institute of Technology and California Institute of Technology. Despite the controversy, an article in the January-February 1997 issue of Mercury magazine, suggested that this discovery was an opportunity for teachers. The article even suggests that there is a possibility of the existence of Martian aliens.

Conclusion

Our study has taken us through the basic concepts of science and began with a look at what science really is and how the scientific method can be applied. The conclusion is that there is no contradiction between being a Christian and being a scientist. We also saw that neither the creation model nor the evolution model could be classified as a theory. Neither is testable or capable of being disproven. Therefore we can only call them models and see how the evidence fits each model.

A study of the principles of chemistry and biology provided a background on the fundamentals which served two purposes. The first was to provide you with a greater appreciation of the creation through understanding the underpinnings of the material and energy world. This shows a little of the greatness of God. Second, the chemical and biological principles are important building block concepts used in both the creation and evolution models. Before we could guide you through the analysis of these models, their biological and chemical components needed to be understood.

In Chapter 5, the principles for the creation model were presented: (1) God created the first life and (2) each species (or kind) was created in the beginning. These species remain distinct; however, within the species variation is explained by natural selection -- a God-given capability for organisms to change in their forms (such as hybrid corn or breed of dog). This might be thought of as a repertoire of genetic potential. The basis for these principles were drawn from Genesis. In can be concluded from the discussion in that chapter, that the creation model was possible and, based on the evidence, the best explanation of our living world.

In Chapter 6, the evolution model was presented in the same light with an examination of the same two areas (i.e., the origin of life and the explanation for diversity). The evolution model attributes the origin of life to chance random assemblage of particles. Regarding the diversity of life, it acknowledges the concept of natural selection as a basis for change, however, the evolution model allows for one species to evolve into a new species. We explained that there is no evidence for this to be found. Darwin, in his original book, declared that there would surely be evidence in the fossil record in the form of transitional forms (one species becoming another), but there is not! We should add that the chapter on evolution was the most difficult to write because of the lack of substance from which we could draw any evolution premises. We examined writings from the disciplines of geology, paleontology, ecology, chemistry, biology, and history. We looked at modern material as well as historical data. What we found were dogmatic (confident) statements about evolution theory, but no substantial evidence.

In this final chapter, we have looked at some questions that come up from time to time. Although we can't say with certainty just how old the earth is, there is definitely no scientific grounds for an old earth. We know the fossil record is good evidence for the creation model. Collecting fossils can be fun and offers us an opportunity to see many different examples of God's creation. Even the existence and extinction of dinosaurs in the past shows evidence for the changes in the earth that we would expect as a result of a catastrophic flood.

The final area we discussed was the recent suggestion that life originated on Mars. It's interesting that the pattern continues of trying to answer questions in ways that deny God's power and plan. It seems so desperate when the answers are found clearly and simply in the Holy Scriptures.

However, we shouldn't be surprised. The Apostle Paul warned that these times would come when he wrote to Timothy:

> For the time will come when men will not put up with sound doctrine. Instead, to suit their own desires, they will gather around them a great number of teachers to say what their itching ears want to hear. They will turn their ears away from the truth and turn aside to myths. (2Tim. 4:3-4)

Discussion Questions

1. Is it hard to imagine man and dinosaur living at the same time? Why? What evidence is there that they did coexist?

2. Do you believe there is or could be intelligent life on other planets? What impact would that have on Christianity?

Bibliography and Additional Reading List

Associated Press. NASA Experts Defend Theory of Life on Mars. (From NEXIS Abstract No. 9300159612) *New York Times*, Sec A, p. 24, Oct 25, 1996.

Baker, S. *Bone of Contention*. Hertfordshire, England, Evangelical Press, 1990.

Bergman, J. Influence of Evolution on Nazi Thought. *Papers of the 1983 National Creation Conference*, pp. 156-160. Richfield , Minnesota, Onesimus Publishing, 1985.

Bram, L. L. and Dickey, N. H. Geology. *Funk and Wagnalls New Encyclopedia* (Vol. 11, p. 272) 1986.

Charig, A. *A New Look at Dinosaurs*. New York, Facts on File, Inc., 1988.

Chiras, D. D. *Biology The Web of Life*. Minneapolis/St. Paul, West Publishing Co., 1993.

Cowen, R. Martian Message - There was Once Ancient Life Here. (From Proquest Abstract No. 04170423) *Christian Science Monitor*, Col 1, p. 1, Aug 8, 1996.

Cowen, R. More Findings About Life on the Red Planet. *Science News,* 151, p. 87, Feb 8, 1997.

Darwin, C. *The Origin of Species.* Great Britain, Penguin, 1987.

Davis, J. D. *Davis Dictionary of the Bible.* Nashville, Broadman Press, 1973.

deBeer, G. (Ed.) *Charles Darwin Thomas Henry Huxley Autobiographies.* Oxford, Oxford University Press, 1983.

Denton, M. *Evolution: A Theory in Crisis.* Bethesda, Maryland, Adler and Adler, 1985.

Gayrard-Valy, Y. *Fossils: Evidence of Vanished Worlds.* New York, Harry M. Abrams, Inc., 1994.

Gish, D. T. *Evolution: The Challenge of the Fossil Record.* El Cajon, California, Creation-Life Publishers, 1986.

Holmes, B. Death Knell for Martian Life. (From Expanded Academic ASAP Database Abstract No. A19110017) *New Scientist,* Vol 152, p. 4, Dec 21, 1996.

Lambert, D. *The Field Guide to Prehistoric Life.* New York, Facts on File Publications, 1985.

Lockwood, J. F. *The Magic of Mars.* (From Expanded Academic ASAP Database Abstract No. A19150141) Mercury, Vol 26, p. 9, Jan-Feb 1997.

Masterton, W. L. and Slowinski, E. J. *Chemical Principles.* Philadelphia, W. B. Saunders Company, 1973.

McCulley, C. By the Book? *Crystal City, etc,* Fall, 1996.

MacFall, R. P. and Wollin, J. C. *Fossils for Amateurs A Handbook for Collectors.* New York, Van Nostrand Reinhold Company, 1972.

Morris, H. M. *Men of Science Men of God.* El Cajon, California, Creation-Life, 1988.

Oakland, R. *The Evidence for Creation.* (Video Tape) Oakland Communications, Inc., 1995.

Petersen, D. R. *Unlocking the Mysteries of Creation.* El Dorado, CA, Creation Research Foundation, 1988.

Pitman, M. *Adam and Evolution.* London, Rider and Company, 1984.

Plummer, C. C. and McGeary, D. *Physical Geology.* Dubuque, William C. Brown Company, 1982.

Strahler, A. N. *Introduction to Physical Geography*. New York, John Wiley & Sons, 1970.

Suzuki, D. T., Griffiths, A. J. F., and Lewontin, R. C. *An Introduction to Genetic Analysis*. San Francisco, W. H. Freeman and Company, 1981.

Tattersall, I. *The Fossil Trail*. New York, Oxford University Press, 1995.

Tenney, M. C. *The Zondervan Bible Dictionary*. Grand Rapids, Zondervan Publishing House, 1967.

Toner, M. Team Offers Compelling Clues. (From Proquest Abstract No. 04165157) Atlanta Constitution, Sec A, Col 1, p. 3, Aug 8, 1996.

Unger, M. F. *Unger Bible Dictionary*. Chicago, Moody Press, 1980.

Vestal, D. *The Doctrine of Creation*. Nashville, Convention Press, 1989.

Whitcomb, J. C. and Morris H. M. *The Genesis Flood*. The Presbyterian and Reformed Publishing Company, 1981.

Wilson, E. Cold Water Thrown on Martian Life Hypothesis. (From Expanded Academic ASAP Database Abstract No. A19130665) *Chemical Engineering News*, Vol 75, p. 8, Jan 6, 1997.

Appendix A

Creation Related Scriptures

Selected Old Testament References to the Creation

Gen. 1:1   In the beginning God created the heavens and the earth.

Gen. 1:7   So God made the expanse and separated the water under the expanse from the water above it. And it was so.

Gen. 1:16   God made two great lights --the greater light to govern the day and the lesser light to govern the night. He also made the stars.

Gen. 1:21   So God created the great creatures of the sea and every living and moving thing with which the water teems, according to their kinds, and every winged bird according to its kind. And God saw that it was good.

Gen. 1:25   God made the wild animals according to their kinds, the livestock according to their kinds, and all the creatures that move along the ground according to their kinds. And God saw that it was good.

Gen. 1:27   So God created man in his own image, in the image of God he created him; male and female he created them.

Gen. 1:31   God saw all that he had made, and it was very good. And there was evening, and there was morning --the sixth day.

Gen. 2:3   And God blessed the seventh day and made it holy, because on it he rested from all the work of creating that he had done.

Gen. 2:4   This is the account of the heavens and the earth when they were created. When the LORD God made the earth and the heavens--

Gen. 2:9   And the LORD God made all kinds of trees grow out of the ground --trees that were pleasing to the eye and good for food. In the middle of the garden were the tree of life and the tree of the knowledge of good and evil.

Gen. 2:22   Then the LORD God made a woman from the rib he had taken out of the man, and he brought her to the man.

Gen. 5:1   This is the written account of Adam's line. When God created man, he made him in the likeness of God.

Gen. 5:2   He created them male and female and blessed them. And when they were created, he called them "man. "

Exod. 20:11   For in six days the LORD made the heavens and the earth, the sea, and all that is in them, but he rested on the seventh day. Therefore the LORD blessed the Sabbath day and made it holy.

Deut. 32:6   Is this the way you repay the LORD, O foolish and unwise people? Is he not your Father, your Creator, who made you and formed you?
Ps. 89:12   You created the north and the south; Tabor and Hermon sing for joy at your name.

Ps. 95:5   The sea is his, for he made it, and his hands formed the dry land.

Ps. 100:3   Know that the LORD is God. It is he who made us, and we are his; we are his people, the sheep of his pasture.

Ps. 102:18   Let this be written for a future generation, that a people not yet created may praise the LORD:

Ps. 139:13   For you created my inmost being; you knit me together in my mother's womb.

Ps. 148:5   Let them praise the name of the LORD, for he commanded and they were created.

Eccl. 3:11   He has made everything beautiful in its time. He has also set eternity in the hearts of men; yet they cannot fathom what God has done from beginning to end.

Isa. 37:16　　"O LORD Almighty, God of Israel, enthroned between the cherubim, you alone are God over all the kingdoms of the earth. You have made heaven and earth.

Isa. 40:26　Lift your eyes and look to the heavens: Who created all these? He who brings out the starry host one by one, and calls them each by name. Because of his great power and mighty strength, not one of them is missing.

Isa. 40:28　Do you not know? Have you not heard? The LORD is the everlasting God, the Creator of the ends of the earth. He will not grow tired or weary, and his understanding no one can fathom.

Isa. 41:20　so that people may see and know, may consider and understand, that the hand of the LORD has done this, that the Holy One of Israel has created it.

Isa. 42:5　This is what God the LORD says-- he who created the heavens and stretched them out, who spread out the earth and all that comes out of it, who gives breath to its people, and life to those who walk on it:

Isa. 43:1　But now, this is what the LORD says-- he who created you, O Jacob, he who formed you, O Israel: "Fear not, for I have redeemed you; I have summoned you by name; you are mine.

Isa. 43:7   everyone who is called by my name, whom I created for my glory, whom I formed and made."

Isa. 45:7   I form the light and create darkness, I bring prosperity and create disaster; I, the LORD, do all these things.

Isa. 45:8   "You heavens above, rain down righteousness; let the clouds shower it down. Let the earth open wide, let salvation spring up, let righteousness grow with it; I, the LORD, have created it.

Isa. 45:12   It is I who made the earth and created mankind upon it. My own hands stretched out the heavens; I marshaled their starry hosts.

Isa. 45:18   For this is what the LORD says-- he who created the heavens, he is God; he who fashioned and made the earth, he founded it; he did not create it to be empty, but formed it to be inhabited-- he says: "I am the LORD, and there is no other.

Jer. 10:12   But God made the earth by his power; he founded the world by his wisdom and stretched out the heavens by his understanding.

Jer. 33:2   "This is what the LORD says, he who made the earth, the LORD who formed it and established it --the LORD is his name: Jer. 51:15 "He made the earth by his power; he founded the world by his wisdom and stretched out the heavens by his understanding.

Amos 5:8   (he who made the Pleiades and Orion, who turns blackness into dawn and darkens day into night, who calls for the waters of the sea and pours them out over the face of the land-- the LORD is his name--

God Spoke to Job About the Creation

Job 38:4-41 4 "Where were you when I laid the earth's foundation? Tell me, if you understand.   5 Who marked off its dimensions? Surely you know! Who stretched a measuring line across it?   6 On what were its footings set, or who laid its cornerstone--   7 while the morning stars sang together and all the angels shouted for joy?   8 "Who shut up the sea behind doors when it burst forth from the womb,   9 when I made the clouds its garment and wrapped it in thick darkness,   10 when I fixed limits for it and set its doors and bars in place,   11 when I said, `This far you may come and no farther; here is where your proud waves halt'?   12 "Have you ever given orders to the morning, or shown the dawn its place,   13 that it might take the earth by the edges and shake the wicked out of it?   14 The earth takes shape like clay under a seal; its features stand out like those of a garment. 15 The wicked are denied their light, and their upraised arm is broken. 16 "Have you journeyed to the springs of the sea or walked in the recesses of the deep? 17 Have the gates of death been shown to you? Have you seen the gates of the shadow of death? 18 Have you comprehended the vast expanses of the earth? Tell me, if you know all this. 19 "What is the way to the abode of light? And where does darkness reside? 20 Can you take them to their places? Do you know the paths to their dwellings? 21 Surely you know, for you were already born! You have lived so many years! 22 "Have you entered the storehouses of the snow or seen the storehouses of the hail, 23 which I reserve for times of trouble, for days of war and battle? 24 What is the way to the place where the lightning is dispersed, or the place where the east winds are scattered over the earth? 25 Who cuts a channel for the torrents of rain, and a path for the thunderstorm, 26 to water a land where no man lives, a desert with no one in it, 27 to satisfy a desolate wasteland and make it sprout with grass? 28 Does the rain have a father? Who fathers the drops of dew? 29 From whose womb comes the ice? Who gives birth to the frost from the heavens 30 when the waters become hard as stone, when the surface of the deep is frozen? 31 "Can you bind the beautiful Pleiades? Can you loose the cords of Orion? 32 Can you bring forth the constellations in their seasons or lead out the Bear with its cubs? 33 Do you know the laws of the heavens? Can you set up [God's] dominion over the earth? 34 "Can you raise your voice to the clouds and cover yourself with a flood of water? 35 Do you send the lightning bolts on their way? Do they report to you, `Here we are'? 36 Who endowed the heart with wisdom or gave understanding to the mind? 37 Who has the wisdom to count the clouds? Who can tip over the water jars of the heavens 38 when the dust becomes hard and the clods of earth stick together? 39 "Do you hunt the prey for the lioness and satisfy the hunger of the lions 40 when they crouch in their dens or lie in wait in a thicket? 41 Who provides food for the raven when its young cry out to God and wander about for lack of food?

Job 39:1-30 1 "Do you know when the mountain goats give birth? Do you watch when the doe bears her fawn? 2 Do you count the months till they bear? Do you know the time they give birth? 3 They crouch down and bring forth their young; their labor pains are ended. 4 Their young thrive and grow strong in the wilds; they leave and do not return. 5 "Who let the wild donkey go free? Who untied his ropes? 6 I gave him the wasteland as his home, the salt flats as his habitat. 7 He laughs at the commotion in the town; he does not hear a driver's shout. 8 He ranges the hills for his pasture and searches for any green thing. 9 "Will the wild ox consent to serve you? Will he stay by your manger at night? 10 Can you hold him to the furrow with a harness? Will he till the valleys behind you? 11 Will you rely on him for his great strength? Will you leave your heavy work to him? 12 Can you trust him to bring in your grain and gather it to your threshing floor? 13 "The wings of the ostrich flap joyfully, but they cannot compare with the pinions and feathers of the stork. 14 She lays her eggs on the ground and lets them warm in the sand, 15 unmindful that a foot may crush them, that some wild animal may trample them. 16 She treats her young harshly, as if they were not hers; she cares not that her labor was in vain, 17 for God did not endow her with wisdom or give her a share of good sense. 18 Yet when she spreads her feathers to run, she laughs at horse and rider. 19 "Do you give the horse his strength or clothe his neck with a flowing mane? 20 Do you make him leap like a locust, striking terror with his proud snorting? 21 He paws fiercely, rejoicing in his strength, and charges into the fray. 22 He laughs at fear, afraid of nothing; he does not shy away from the sword. 23 The quiver rattles against his side, along with the flashing spear and lance. 24 In frenzied excitement he eats up the ground; he cannot stand still when the trumpet sounds. 25 At the blast of the trumpet he snorts, `Aha!' He catches the scent of battle from afar, the shout of commanders and the battle cry. 26 "Does the hawk take flight by your wisdom and spread his wings toward the south? 27 Does the eagle soar at your command and build his nest on high? 28 He dwells on a cliff and stays there at night; a rocky crag is his stronghold. 29 From there he seeks out his food; his eyes detect it from afar. 30 His young ones feast on blood, and where the slain are, there is he."

Selected New Testament References to the Creation

Acts 4:24   When they heard this, they raised their voices together in prayer to God. "Sovereign Lord," they said, "you made the heaven and the earth and the sea, and everything in them.

Acts 14:15   "Men, why are you doing this? We too are only men, human like you. We are bringing you good news, telling you to turn from these worthless things to the living God, who made heaven and earth and sea and everything in them.

Acts 17:24   "The God who made the world and everything in it is the Lord of heaven and earth and does not live in temples built by hands.

Rom. 1:20   For since the creation of the world God's invisible qualities --his eternal power and divine nature --have been clearly seen, being understood from what has been made, so that men are without excuse.

Eph. 2:10   For we are God's workmanship, created in Christ Jesus to do good works, which God prepared in advance for us to do.

Eph. 3:9   and to make plain to everyone the administration of this mystery, which for ages past was kept hidden in God, who created all things.

Col. 1:16   For by him all things were created: things in heaven and on earth, visible and invisible, whether thrones or powers or rulers or authorities; all things were created by him and for him.

1Tim. 4:4   For everything God created is good, and nothing is to be rejected if it is received with thanksgiving,

Rev. 4:11   "You are worthy, our Lord and God, to receive glory and honor and power, for you created all things, and by your will they were created and have their being."

Appendix B

The Plan of Salvation

The Good News

God's desire is for us to have eternal life with him. That's why we were created. We can have that eternal life beginning now.

John 3:16 "For God so loved the world that he gave his one and only Son, that whoever believes in him shall not perish but have eternal life.

John 10:10 The thief comes only to steal and kill and destroy; I have come that they may have life, and have it to the full.

Rom. 5:1 Therefore, since we have been justified through faith, we have peace with God through our Lord Jesus Christ,

The problem started with the very first man's rebellion. Having eternal life is not automatic. You have a different nature, and you are separated from Him and Life by your own sin.

Rom. 3:23 for all have sinned and fall short of the glory of God,

God provided the way for the fellowship to be restored by sending His Son to pay for our sins.

Rom. 6:23 For the wages of sin is death, but the gift of God is eternal life in Christ Jesus our Lord.

Rom. 5:8 But God demonstrates his own love for us in this: While we were still sinners, Christ died for us.

God's way is the only way.

> John 14:6   Jesus answered, "I am the way and the truth and the life. No one comes to the Father except through me.

> Eph. 2:8-9   For it is by grace you have been saved, through faith --and this not from yourselves, it is the gift of God-- not by works, so that no one can boast.

What can you do? Believe and receive.

> John 1:12   Yet to all who received him, to those who believed in his name, he gave the right to become children of God--

> Rom. 10:9   That if you confess with your mouth, "Jesus is Lord," and believe in your heart that God raised him from the dead, you will be saved.

> Rom. 10:10   For it is with your heart that you believe and are justified, and it is with your mouth that you confess and are saved.

> Rom. 10:13   for, "Everyone who calls on the name of the Lord will be saved."

Pray and ask God to save you.

> God, I know I am a sinner.
> I believe Jesus died for my sins.
> Right now I repent and turn from my sins.
> I open the door of my heart and receive Jesus as my personal Savior.
> I commit myself to follow you.
> Thank you for saving me. Amen.

Jesus said: Here I am! I stand
at the door and knock. If
anyone hears my voice and
opens the door, I will come in
and eat with him, and he with
me. Rev. 3:20

2Cor. 5:17   Therefore, if
anyone is in Christ, he is a new
creation; the old has gone, the
new has come!

The Authors

Elizabeth J. Ridlon is a former college teacher. She served as an adjunct faculty member at Belleville Area College in Illinois from 1990 to 1995, teaching General Biology, Human Biology, and Anatomy and Physiology. Prior to that, she taught biology at the University of Maryland, European Division. Mrs. Ridlon earned a Bachelor of Science Degree in Microbiology from Indiana University, a Master of Arts Degree in Biology from the University of Nebraska, and a Secondary Teachers Certificate from Southern Illinois University at Edwardsville. Her masters thesis was a study of seasonal changes in the blood biochemistry of the Western Plains Garter Snake. She is a member of First Baptist Church of O'Fallon, Illinois where she is active in various ministries.

Robert W. Ridlon, Jr. is a consultant for an information systems development firm. He has been an adjunct faculty member at Belleville Area College, Illinois since 1991, and teaches information systems theory; and systems analysis and design. Mr. Ridlon earned a Bachelor of Arts degree in Biological Sciences from Indiana University and a Master of Science Degree from the Air Force Institute of Technology. His masters thesis was a study on organizational behavior. Mr. Ridlon has several scientific and technical publications to his name in the areas of biochemistry, cell biology, and information systems. Mr. Ridlon is a member and an ordained Deacon at First Baptist Church of O'Fallon, Illinois.

The Ridlons have one son, Robert W. Ridlon, III, who attends college. As a family, they have traveled extensively on four continents and enjoy photography and history (especially Bible history).

To order additional copies of *Understanding the Origin and Diversity of Life*, write to:

Jordan Hall Publishing
P.O. Box 364
Troy, Illinois 62294-0364